ALSO BY JAY INGRAM

THE FUTURE OF US

OF US

The Science of What We'll Eat, Where We'll Live, and Who We'll Be

JAY INGRAM

Published by Simon & Schuster

NEW YORK   LONDON   TORONTO
SYDNEY   NEW DELHI

A Division of Simon & Schuster, LLC
166 King Street East, Suite 300
Toronto, Ontario M5A 1J3

This Simon & Schuster Canada edition October 2024

SIMON & SCHUSTER CANADA and colophon are trademarks
of Simon & Schuster, LLC

Simon & Schuster: Celebrating 100 Years of Publishing in 2024

For information about special discounts for bulk purchases,
please contact Simon & Schuster Special Sales at 1-800-268-3216
or CustomerService@simonandschuster.ca.

Manufactured in the United States of America

1   3   5   7   9   10   8   6   4   2

Library and Archives Canada Cataloguing in Publication

Title: The future of us : the science of what we'll eat,
where we'll live, and who we'll be / Jay Ingram.
Names: Ingram, Jay, author.
Description: Simon & Schuster Canada edition. | Published in hardcover in 2023. |
Includes bibliographical references and index.
Identifiers: Canadiana 20240361490 | ISBN 9781668081150 (softcover)
Subjects: LCSH: Forecasting—Study and teaching—Popular works. |
LCSH: Science—Popular works.
Classification: LCC CB161 .I54 2024 | DDC 303.4909/05—dc23

ISBN 978-1-6680-8115-0
ISBN 978-1-6680-0334-3 (hc)
ISBN 978-1-6680-0335-0 (ebook)

Contents

To MA for, among many other things,
her infectious spirit of adventure

Prediction is very difficult, especially if it's about the future.
—Variously attributed to physicist Niels Bohr or
New York Yankees' philosopher Lawrence Peter Berra

THE
FUTURE
OF US

The Fog of the Future

The 2020s are already an astonishing decade of technological innovation. On December 14, 2022, scientists at the Lawrence Livermore National Laboratory in California announced they had achieved nuclear fusion. Instead of splitting atoms (fission) to generate nuclear power, atoms are forced together (fusion). This was a critical breakthrough, because for a short time, more energy was generated by the fusion technology than had been put in, obviously a crucial milestone. For half a century, science had been working toward this major breakthrough; it could take another half century to turn this single demonstration into a technology that will be applicable globally, but because nuclear fusion generates no carbon dioxide or any nuclear waste, it would be an ideal technology for reducing carbon emissions.

That's just one example. Just three months earlier, on September 13, 2022, in the absence of eggs, sperm, and a womb, two independent teams of scientists were able to create mouse *embryos*. While they were not quite complete embryos—they stopped developing at about eight and a half days instead of the normal twenty—these lab-grown specimens were on track to becoming normal embryos.

The scientists began with stem cells. In a living animal these cells begin life with no cellular identity—they aren't yet liver cells, blood cells, or skin cells. But then they transform into specialized cells and populate our bodies with unique tissues and organs.

That's the normal version of embryonic development. But the

scientists on these two teams took this stem cell versatility in a whole new direction: they were able to coax a group of mouse stem cells to begin to develop an embryo simply by putting the cells in a set of spinning glass vials inside an incubator, thus creating an artificial womb designed to mimic the circulation of blood and nutrients in the uterus.

While the percentage of normally formed embryos was small, just one-half of 1 percent, *still* they were viable, with beating hearts, normally shaped brains (including the beginnings of defined brain regions), and intestines—confirmation that the programming of the embryo is internal to the stem cells themselves. The uterus is a habitat, not the controller. Of all the quotes in response to these experiments, Case Western Reserve University biologist Paul Tesar's "It's a pretty wild and remarkable time" seems the most appropriate.[1]

One of the teams argued that this breakthrough opens up a new approach to studying embryonic development: it happens right there, before your eyes, in transparent glassware. It will be possible to see exactly when and where embryonic development goes wrong. With some tinkering, it should one day be possible to carry the embryos to term. A mature mouse embryo created from stem cells would be a dramatic scientific achievement.

Immediately after the release of the scientific paper, there were thoughts (doubts? fears?) about doing the same thing with humans, which would provide unparalleled insights into early fetal development.

And a third example: On October 12, 2022, an article was published in the journal *Neuron* that did indeed involve neurons—nerve cells—from both mice and humans (the human neurons were cultured from stem cells, the mouse version from mouse brains).[2] These neurons, hundreds of thousands of them, were hooked up to a set of electrodes, facilitating electrical communication among them. Amazingly this half-biology, half-technology setup (dubbed Dishbrain) could be programmed to play a lab version of the old video game *Pong*. The neurons sped up the whole circuit and played *Pong* really well. The setup also revealed that human neurons were better at the game than mouse neurons. The research was a step forward in combining an artificial brain with a real one.

Tech Never Stands Alone

These are just some recent examples of how technology is accelerating the possibilities of real-world application—and complications. There isn't a topic in this book that isn't both moving forward dramatically while at the same time creating ethical and equity issues.

For example, on November 26, 2018, the now disgraced Chinese biophysicist He Jiankui used CRISPR gene-editing technologies to edit the genomes of twins while they were still in the womb. Even though he was jailed and fined for it, can we be certain that someone else won't try to use human stem cells to develop a human embryo?[3]

It's impossible to predict which of these examples of pioneering science will take off, where they might go, or how far. In mid-2021, stocks of Beyond Meat, the giant producer of plant-based meat, were trading at US$150. It looked like a dramatic step forward for plant-based products and the green values associated with them. One year later, the stock had dropped by US$75. It had looked like a winner—and then it wasn't. And who knows? Since then that market has quietly become crowded.

But one thing is for certain: there will be technologies with enormous impact. One of the most pressing issues facing these new technologies will be "Who benefits?"

Future Thinking

We will need to be able to think more effectively about the future, something that we don't routinely do today, at least on the long-term time scales we need. Thinking about the future is not even an integral part of human nature, and has waxed and waned with the centuries. Medieval scholars believed that life in the 1300s and 1400s consisted of endless cycles: good crop years and bad, pestilence and health, good kings and bad. Only the Bible promised a future, and an apocalyptic one at that.

But today, when even short-term forecasts promise that most of the jobs we're familiar with now will not exist in the future, we have to try to account for what will happen. If we don't, we may end up the way the horses did. Max Tegmark, in his book *Life 3.0*, asked his readers to imagine two horses looking at an early automobile in the year 1900 and pondering their future:

"I'm worried about technological unemployment."

"Neigh neigh don't be a Luddite: our ancestors said the same thing when steam engines took our industrial jobs and trains took our jobs pulling stagecoaches. But we have more jobs than ever today, and they're better too: I'd much rather pull a light carriage through town than spend all day walking in circles to power a stupid mine-shaft pump."

"But what if this internal combustion engine thing really takes off?"

"I'm sure there'll be new jobs for horses that we haven't yet imagined. That's what's always happened before, like with the invention of the wheel and the plough."[4]

Tegmark points out that in the end, the horse population went from 26 million in 1915 to 3 million by 1960.

Should you be optimistic or pessimistic about the future? Wherever you land on that question will be more a reflection of your personal outlook than any compilation of data and trends. Protecting the Earth against climate change has seemed hopeless at times—and still does to many—but acknowledgment of the issue and support for action are gaining traction. Optimists like Stanford sociologist Rob Reich imagined this future on the Singularity Radio podcast *The Feedback Loop*: "We're leaving behind an era of leaving it to the technologists alone to fix the problems . . . now we are going to have an array of countervailing forces and opportunities for people to think about the great digital revolution and its extraordinary promises and how to ensure it goes in a socially positive direction, democratically supportive direction."[5]

While Reich is specifically targeting digital technologies, let's hope his thinking applies to new technologies generally. The best advice I can give? Prepare for the future with Arthur C. Clarke's words in mind: "Only if what I tell you appears absolutely unbelievable, have we any chance of visualizing the future as it will really happen."

PART I

All About Us

CHAPTER 1

Our Future Selves

Our thoughts about the future naturally focus on ourselves. But thoughts about the imagined self are usually short-term and centered on what we will be doing, not what we will look like. Our bodies have changed in the past; what direction will they take in the future, hundreds or thousands of years from now? First, set aside the possibility that every human in that distant future will be some kind of hybrid of flesh-and-blood and technology (in other words, don't be swayed by what's talked about most these days) and think about the human body on its own. It has evolved dramatically over the last 200,000 years en route to becoming a new species, but what about from this day forward? Read on.

"Where are we going? Life, the timeless and mysterious gift, is still evolving. What wonders or terrors does evolution hold in store for us in the next ten thousand years? In a million?"

These portentous words opened the episode titled "The Sixth Finger" in the 1963 season of the science fiction TV show *The Outer Limits*.[1] In this episode, a scientist in Wales discovers a way of accelerating human evolution. To expiate his guilt over his invention of a super atomic bomb, he proposes to create a man with the intellectual abilities of future humans—a man who might save the world. The man he chooses is a disgruntled coal miner named Wilhelm Griffith, played by David McCallum (better known for his roles as Eric Ashley-Pitt in *The Great Escape* and Illya Kuryakin in *The Man from U.N.C.L.E.*). Griffith agrees to undergo the accelerated mutation treatment, which propels him far into the future. By the time he's evolved twenty thousand years forward, his cerebral cortex has swelled and a sixth finger has grown on each hand (more about this later). By the time he's reached a million years hence, his skull is enormous, as is his forehead, and his ears are large and pointy. But the rest of him hasn't much changed, other than that finger.

Looking Forward to Big Heads

This episode of *The Outer Limits* is a striking example of the common belief that as we evolve, the most significant change to the human body will be the growth of our brains and thus the size of our heads. The theory's persistence also owes a great debt to one of the most famous depictions of the history—actually, the *prehistory*—of humans.

The March of Progress, originally titled "The Road to Homo Sapiens," first appeared in a book called *Early Man*, published by Time-Life Books in 1965. The four-and-a-half-page spread depicted what was thought at the time to be the evolution of humans. Starting with a small, ancient four-legged primate, the spread revealed six individuals walking single file from left to right, ending with modern humans.[2]

The March of Progress has not aged well, scientifically or socially. Besides the sole focus on "man," the science of half a century ago has been overhauled by a multitude of fossil finds. Human evolution can no longer be conceived as anything like a straight line; instead, it is a weaving together of multiple species, some of whom lived at the same time in the same place. The fact that we have a few percent Neanderthal DNA in our genome is solid proof of that intermingling.

But even if the March of Progress is inaccurate, if not inappropriate, it continues to be an enduring representation of human evolution. Two features particularly stand out: first, the move from four-legged locomotion to two; and second, the growth of the brain. If you were to extrapolate the March of Progress well into the future, it seems reasonable that the increase in brain size would be the trend most likely to continue: bigger brain, bigger head, higher forehead (fitting nicely with the ridiculous notion that high foreheads, that is, receding hairlines, signify intelligence).

However, is it reasonable to predict that the human brain will continue to evolve and grow into the future? Are humans even still evolving? There are data that relate to both these questions, but the view they present of the future is cloudy.

First, the enlargement of the brain. Over a few million years, our brain size has tripled. Before our genus, *Homo*, appeared, the last of the Australopithecines, *Australopithecus afarensis*, which lived about three million years ago, walked upright (if not as effortlessly as we do). Males stood five feet tall, females less than four feet, and the volume of their brains was half a liter (about a pint). Arriving on the scene about a million years later, *Homo erectus*—which, while not the same species as we are, in many ways was definitively human—was close to six feet tall in some cases, proficiently bipedal, and had a brain volume that grew over time, reaching a liter (a quart) or slightly more.

More recently, brain size leveled out or even dipped slightly: today, the human brain averages about 1,350 cubic centimeters or 82 cubic inches, although a healthy brain can be a couple of hundred cubic centimeters more or less in either direction. But there's no doubt that over this time span, our brain size has increased steadily, suggesting a trend that could continue.

Missing from that brief list of species are the Neanderthals, who, despite going extinct about forty thousand years ago in Europe, had brains at least as big or bigger than ours, about 1,500 cubic centimeters (91.5 cubic inches). Even though they were not our direct ancestors, the Neanderthals were very humanlike and a living demonstration that brain size alone does not guarantee evolutionary success.

That's not all. While the evolution of larger brains holds true over the grand sweep of human evolution, ongoing research reveals the process to be anything but inevitable. In the last twenty thousand years—and modern humans were the only hominin species left by that time—the male brain has lost about 150 cubic centimeters or 9 cubic inches. That's pretty close to the volume of a tennis ball, which brings a 1,500-cubic-centimeter or 91.5-cubic-inch brain down to the 1,350 cubic centimeters or 82 cubic inches I mentioned earlier.[3] Female humans, whose brains are marginally smaller, housed as they are in smaller bodies on average, have shrunk by the same proportion. This is not a particularly helpful comparison: a hole in your head 150 cubic centimeters or 9 cubic inches (about the size of a tennis ball) seems catastrophic, although imagining that loss spread evenly over the entire cerebral cortex does lessen the blow. Still, it's a dramatic reversal from the general growth trend. In fact, if that trend were to continue for another twenty thousand years, our brains would revert back to the size of *Homo erectus*'s.

In another study, measurements of the skulls of white Americans born between the 1820s and 1980s showed that over a century and a half the skull evolved to be higher and longer, although slightly narrower. Despite the narrowing, the inside volume increased in an amount approximately equal to—ready for this?—the volume of a tennis ball. (That study included only white Americans because they represented the bulk of the available samples, underlining the fact that more widely representative and inclusive studies of this kind are much needed.)[4]

There are many data sets; the detailed volumes are less important than the trends, which show that as our distant ancestors evolved, brains trended larger. But a few hundred thousand years ago, the brain reached a maximum size, and that size has fluctuated ever since. So what appears to be a brain growth trend turns out on closer inspection to be illusory. And frankly, head size at birth is constrained by the size of the human birth canal, so balloon-headed future humans would have to postpone a huge amount of head growth until after they were born.

Head and Shoulders, Knees and Toes

What accounts for the ups and downs of human head (and brain) size? Some clues might lie in the changes in the rest of the body, particularly stature. Like brain size, the height of our ancestors has zigged and zagged.

Nutrition, and by association agriculture, hunting, and climate, have also played a major role.

Roughly ten thousand years ago, the Agricultural Revolution led to dramatic increases in population, but paradoxically depressed height. Several possible reasons have been suggested: occasional crop failures impacted nutrition and growth; people living in closer proximity promoted the spread of growth-slowing diseases; and even reliance on a plant diet rather than the mixed hunter-gatherer diet could all have played a role.

Jumping ahead to the Middle Ages, say 900–1000 CE, humans reached heights that weren't matched again until the nineteenth century! Again, cold temperatures in the 1600s and 1700s are thought to have diminished agricultural output; only in the last two hundred years have we gained stature again.

Today in most places in the world, we are as tall as we have ever been, but even so there are oddities in the data. For instance, in the early nineteenth century the Dutch were among the shortest people in Europe, averaging only 5 feet 5 inches. But by the late twentieth century, they had gained an amazing 7 inches in average height, so that Dutch men averaged 6 feet tall. Yet, since 1980, the Dutch have not only slowed their increasing growth but have lost about half an inch. It's not clear why: changes in diet might be one reason. Still, the average Dutch man is about 3 inches taller than the average American man—over the past few decades Americans have become heavier, not taller. Interesting to note that in the early 1800s, the Cheyenne were taller than the average American today.[5]

These ups and downs are more generally due to environmental and not genetic effects (although the rapid rise of the tall Dutch was apparently related to tall parents). It's not that genetics is irrelevant; it's that environmental changes seem to have exerted more profound effects than genes. That being the case, predicting the shape of future humans, whether in brain size or height, is extremely difficult.

But even if genetics seems to play a secondary role to the environment today, the rapid growth of the brain and the body that happened early in human evolution was very powerful—could that process assert itself again? It could if there were some dramatic environmental change—catastrophic climate change might be one—but we wouldn't see significant changes in human appearance for thousands of years.

Are We Still Evolving?

Some scientists have claimed that human evolution, having brought us to where we are today—where we control our life circumstances more than the environment does—is now over. For instance, many childhood infections that might have been fatal are now controlled by antibiotics, so any advantage conferred by genetic resistance against the organisms causing those diseases is irrelevant and such genes won't continue to spread throughout the population. The unprecedented mobility of humans also runs counter to the past, when isolated populations would evolve their own unique sets of genes tailored to their specific environment. Now mixing prevails.

Evolution is dependent on a constant supply of mutant genes, a tiny handful of which might confer an advantage that could enhance reproduction. The "choice" of which genes are beneficial is forced by the environment—that's natural selection. There's no reason to think that humans have stopped providing plenty of genetic variation, but is any of that variation being selected and therefore becoming established in the genome? Yes, there are examples of exactly that.

Mere thousands of years ago, resistance to the deadly disease anthrax was one such evolutionary change. Human genes conferring that resistance (at least, partial resistance) only started to take hold in the human genome after early humans started to hunt ruminants. These were the genes that normally coded for the receptor for the anthrax toxin, the molecule that facilitates its entry into our cells. Mutant versions of that gene blocked entry of the toxin and saved lives. These mutants got a second boost around the advent of agriculture, especially when we domesticated cattle and sheep, the prime anthrax bacterial carriers. Environmental changes triggering genetic changes—that's what evolution is all about.

It's almost the same thing with the genes that allow humans to digest lactose beyond infancy. About 35 percent of the world population can digest lactose, most of those being in Europe, especially Northern Europe, the Middle East, and some parts of Africa. It might seem as if the advent of dairy farming provided a benefit to the genes that permit lactose digestion and the people who happened to have those genes reproduced more effectively, allowing those genes to spread throughout those populations. Yet there are still puzzles about the exact sequence of events, particularly because in some parts of the world where a lactose gene is common, there was a lag of a few thousand years between the advent of dairy farming and the establishment of the gene in the population. Astute dairy companies today appeal to both those with and without these genes by selling their usual milk *and* a lactose-free version. If both are available there's no advantage to having a lactose gene and that in turn might slow down its spread (although it's not like lactose-free milk is commonly available).

Some researchers argue that culture can also determine how the human species evolves. What if changing cultural practices were the first step in giving an advantage to those with the right genetics, with the resulting biological evolution (the spread of the genes in the population) following thousands of years later? The power of cultural evolution is that it spreads much, much faster than biology.[6]

Even if/when powered by cultural evolution, changes in the human genome are unlikely in the near term to lead to the kind of fantastical changes pictured in science fiction—the humungous brain or huge saucer eyes. Yes, change is ongoing, although subtle and sometimes difficult to interpret.

The most recent research, while controversial, identified more than *seven hundred* traits controlled by multiple genes where pressure from the environment has been creating evolutionary change over the last two thousand to three thousand years. (They actually looked much further back, in the ballpark of one hundred thousand years, although the rarity and quality of DNA recovered from that time limits what can be concluded about it.)

It sounds simple: if a gene becomes more common over time, it is being selected and therefore must hold some advantage for those it inhabits. In this case, the controversy surrounds the reliability of the statistical methods: traits that are controlled by multiple genes are much more difficult to track. But if these recent stats withstand scrutiny, the data is wide-ranging and sometimes puzzling.

For instance, genes associated with schizophrenia and attention-deficit hyperactivity disorder (ADHD) seem to be on the rise, and their advantage isn't as immediately clear as, say, a gene that confers resistance to a common disease. Similarly, increases in genes associated with anorexia nervosa, inflammatory bowel disease, and skin cancer are puzzling because they compromise survival—and therefore reproduction—so shouldn't be spreading through the genome. But they are. The only reasonable explanation is that there are somehow undiscovered side benefits to the presence of these genes, ensuring that on balance they promote survival. Possibly at a time when hygiene wasn't nearly what it is today, a highly active immune system would have been protective, but today that activity is turned inward, causing bowel disease.

So there's solid evidence that a host of genes are under selection pressure, either being promoted or demoted. Yes, we are still evolving, but in ways where often the survival benefit is not easily apparent.

Michael Lynch, a biologist at Indiana University, argues that there's another trend: our DNA is still mutating, but because we exert so much control over our environment, many mutations that might have had a negative, even fatal, effect in earlier times now seem inconsequential because of compensatory medical and social advances. Lynch argues that such mutations will accumulate, leading to a gradual deterioration of the human genome. He is specifically concerned that because "the brain is a particularly large mutational target," some of these mutations will eventually demand ramped-up neurological responses. [7]

What's the Perfect Number of Fingers?

In *The Outer Limits* episode "The Sixth Finger," the extra digit of the title was, significantly, the only part of Wilhelm's body, other than his enormous head, that appeared to have changed. The suggestion is that somewhere in the human genome there are genes that, if they could be turned on, would run a genetic program to build a sixth finger. This is in contrast to the genetic program for growing a tail—we lost those genes so long ago that there's really no chance they could reassemble. (We do sprout a tail very early in gestation, but it doesn't connect to the spinal cord, has none of its own bones, and so is really at best a pseudo-tail. Besides, what good would it be?)

Wilhelm's sixth finger was an extra baby finger. And though his future self proved to be a brilliant pianist, it wasn't clear that ability had been enhanced by having an extra finger on each hand. Such so-called super-numerary digits occur maybe once in every several hundred births (the statistics are all over the map); in humans, at least, they're usually viewed as a defect and are commonly surgically removed.

Theorist Mark Changizi of 2AI labs used the dimensions of hands and fingers to calculate the ideal number of fingers as 4.78, which rounds up nicely to five on each hand. The evolutionary evidence shows that some of the first animals to emerge onto the land 350 million years ago experimented with different numbers of digits, even up to eight per hand/foot, but over time the number settled down to five. So if five fingers really is the ideal number, surely a sixth couldn't be of much use.[8]

However, a cool study published in *Nature Communications* in 2019 revealed that while extra fingers may be rare, that doesn't mean they can't be advantageous.[9] An international group of scientists studied a mother and her seventeen-year-old son, both of whom had an extra forefinger. They found that not only did these extra fingers have a complete set of their own nerves and muscles, and so were capable of normal finger movements, but each had its own territory in the brain devoted to those movements. There was nothing inferior about these extra fingers. Not only that, these two individuals were able to play a video game using the six fingers on one hand while five-fingered individuals were forced to use all five on one hand and a single finger from the other. Not that the enhanced ability to play a lab video game with one hand represents any sort of evolutionary advantage (unless video game playing becomes a matter of survival), but these experiments did show that at least in certain specific lab circumstances, an extra finger was advantageous.

Most evolutionists, at least at this point, would rate the chances of humans evolving a sixth digit pretty low, but that doesn't deny the possibility of adding a *prosthetic* extra digit. A recent art/science collaboration in England called the Third Thumb Project enlisted people to wear an additional thumb positioned across the hand from the existing one.[10] Just as in the study of the six-fingered mother and her son, the scientists were able to record the brain activity associated with this fake thumb. The more time people spent wearing the thumb, the better the coordination they achieved: they were able to pick up and handle a wineglass or build a tower of wooden pieces while occupied by a math problem. The more time they spent wearing the thumb, the more it felt a part of them. The Third Thumb Project is just a tiny beginning, but it signals that there are possibilities in the future of the human body that might skirt the limitations of evolution as we've experienced it since our beginnings. Our five-fingered hands are so far the end result of millions of years of evolutionary experimentation, but the demonstration that a sixth digit can be fully functional, even advantageous, and a prosthetic thumb that feels as if it's one's own, suggests a future where a lost digit can be completely replaced, whether by a prosthetic or by turning on the genes to grow that sixth finger. If this seems otherworldly, consider that research groups at the University of Freiburg and Imperial College London are investigating the possibility of extra prosthetics, whether a thumb or even an arm.[11] They've discovered that there are "surplus" neural signals involved in movements that could be co-opted (the researchers hope) to move that prosthetic.

Still, replacement, not addition, seems most likely at this point, especially in the absence of a demonstrable need. However, these ideas do suggest that when thinking about the future of the human body, limiting those thoughts to flesh and blood might be missing the real story.

CHAPTER 2

Restoration Hardware

Is your body still the made-to-order standard issue with which you were born? If you have fillings or tooth implants, a pacemaker, tattoos, piercings, an artificial hip or knee or ankle, or artificial lenses or heart valves, then no, it isn't. And that's not counting detachable additions, like glasses or hearing aids. A couple in that list are aesthetic, but the others are upgrades to physical health and well-being. The need for such artificial enhancements has always existed, even when millennia ago people lived much shorter lives, thus reducing the window for accidents or age-related wear-and-tear. It's really only in the last century and a half, and especially the last fifty years, that there have been dramatic improvements in the technology to repair and restore the human body.

While it's true that over millions of years, evolution by natural selection has fashioned our species, its stutter-step inconsistency over the past few millennia means it's unlikely any of us will see substantial change in the appearance of humans in our lifetimes. But that doesn't mean humans are not going to change, maybe dramatically and soon—it's just that we will bring about the change with technology, rather than passively waiting for nature.

A Peg for a Leg, a Hook for a Hand

I'm sure the urge to tinker with the body is ancient. It must go back tens of thousands of years—doesn't it seem reasonable that a human hunter or even a Neanderthal who lost a few fingers or a hand in a confrontation with a cave bear would make some effort to cope with the loss? Create a prosthetic hand by tying a stick with a branched tip to his arm? Walking sticks for sure. A young boy who lived on the island of Borneo thirty-one thousand years ago had his leg amputated when he was around ten or twelve years old and survived another eight or ten years. If humans were skilled enough then to perform an amputation and prevent fatal postsurgical infection, then surely the follow-up of a sport-surgery prosthesis of some kind would have occurred to them. It's unfortunate that given the very short list of materials they could work with, even some sort of primitive artificial lower leg or foot would likely not be preserved for us to find it. But I'm convinced they existed.

Much closer to the present, say three thousand years ago, there's dramatic evidence of skilled replacement of lost body parts. What is sometimes called the "Cairo Toe" is one of the best. This artificial big toe was fitted to the right foot of the mummified body of Tabaketenmut, a priest's daughter who died somewhere between the ages of fifty and sixty. The toe is artfully made of wood and leather, and carefully shaped to fit Tabaketenmut's right foot. The toe had a built-in hinge, presumably to allow it to articulate with the foot. That the wound had time to heal completely means Tabaketenmut had apparently lost that toe long before her death, and the amount of wear on the artificial appendage strongly suggests she had worn it a lot, which resolves a potential ambiguity: if there were no signs of wear it might have been added to her body after death, for mummification.[1]

An artificial replacement for the big toe is a challenging piece of technology: it must provide stability when a person is standing and be able to withstand the force exerted on it when they push off to take a step forward. There is an even older Egyptian toe fabricated from linen, plaster, and glue, and though it too has signs of wear, it has no hinge, suggesting it might have been more cosmetic than functional.

These Egyptian prostheses were way ahead of their time. For centuries after, craftsmanship (as far as we can tell from what's been found) took a back seat to simple practicality: a peg for a leg, a hook for a hand (Long John Silver and Captain Hook come to mind, underlining the hazards of piracy). Accounts of more elaborate prosthetic hands and legs, many of them simply oft-told stories, are impossible to verify. As the Industrial Revolution ushered in machinery, prosthetic replacement became more intricate and somewhat more useful, although until quite recently the materials were heavy and cumbersome, thus compromising comfort and functionality.

Prosthetics have now advanced dramatically, with better, lighter, more flexible materials; more human-centered designs; and dramatic advances in usability. Their use has expanded beyond the replacement of lost limbs. Two modern examples illustrate both the distance we've come—and the distance yet to go.

The Heart: Tissue or Technology?

Heart failure is the leading cause of death in the world. Globally 26 million people suffer from heart failure, and those numbers are anticipated to rise with an aging and growing population. If diseased hearts aren't somehow repaired, assisted, or replaced, death is the result. Replacement is challenging: the heart typically beats more than 100,000 times daily, about 40 million times a year.

The first spectacular advancements in replacement technology were heart transplants. In 1967, Louis Washkansky in South Africa became the first recipient. His surgeon, Christiaan Barnard, became a global celebrity overnight. *Newsweek* called the event "the opening of a new era in medicine, an era as significant as the age of the atom."[2] But it was controversial, too: ethicists argued that Barnard had told Washkansky and his family that the operation had an 80 percent chance of success, which was a gross exaggeration. Indeed, Washkansky lived only eight days with the donor heart. Three days after Barnard performed his transplant, Adrian Kantrowitz performed a transplant on an infant in the US. It, too, died soon after. Barnard grabbed the headlines again by transplanting a heart into Philip Blaiberg, who was a sensation, living nearly six hundred days before succumbing to the multiple impacts of organ rejection.

Today, heart transplants are much more routine and much more successful—five thousand are performed annually around the world. The problem is that there are never enough hearts *available* for transplant. Compare those five thousand operations with the twenty-six million individuals worldwide suffering from heart failure. Obviously, the vast majority never moves to the front of the line and you can be sure that in some parts of the world there isn't even a line to join. This creates a difficult medical situation: by the time a heart is available, many recipients will have declined to the point of being gravely ill and won't survive long even with a new, healthy heart. Despite that hazard, the numbers these days aren't bad: in the United States, chances are pretty good that a recipient will live more than a decade with a transplanted heart.

But the supply problem doesn't have any obvious solutions. In the United States the number of eligible donors is around 2.6 million. But a series of challenges produces sharp reductions in donors: many do not die in the hospital and are therefore inconveniently located for donation; not all potential donors are declared brain dead (and therefore eligible) in time; not all have registered as donors; and not every heart that survives all these challenges is transplanted successfully. And that list covers only a few of the issues. How compromised is the recipient's heart? Does the donor tissue match the recipient's? The number of eligible donors ends up being around three thousand.

You might wait years, you might get a heart in a day or two—there's no way of predicting. But the system is under stress—three thousand patients might get a heart, but thousands more will still be waiting.[3]

Multiple efforts have also been made to create an artificial heart—no flesh and blood involved. Incremental improvements had been going on for decades before the first implantation in 1969. Forty-seven-year-old Haskell Karp from Illinois lived with an artificial heart while waiting for a heart transplant, which took place sixty-four hours later. Karp died two days after. Although the experiment could hardly be called a success, it demonstrated that an artificial heart could keep a person alive, however fleetingly.

The next artificial heart recipient was Barney Clark, a Seattle dentist, thirteen years later.* Clark received a heart called the Jarvik 7, named after American scientist Robert Jarvik, one of the collaborators in its invention. Clark lived for 112 days, and while he had some good moments, it was a very difficult time.

Today the emphasis has shifted to partial, not complete, artificial hearts. So-called ventricular assist devices are being used both as temporary bridges for those on the transplant waiting list ("bridge to transplant") or even as permanent devices to hopefully extend the patient's life ("destination therapy"). These devices assist the diseased heart rather than replacing it, and they're having an impact: a study published at the time of writing by an international team of cardiologists in the *Journal of the American Medical Association* found that survival times on the heart transplant waiting list were improving faster than survival times once the heart is in place![4]

However, the complete artificial heart has not been forgotten, and the technology driving the research is extraordinary. A company called BiVACOR in Huntington Beach, California, is right at the leading edge. In a ninety-day trial study, a cow with a BiVACOR heart remained healthy and active and continued gaining weight. Hearts have now been implanted in both cattle and sheep.

The BiVACOR heart doesn't look like a heart. In fact, it doesn't even beat. It has one moving part: a spinning disc. And there are no valves (and therefore no pulse). Instead, a disc floats in a magnetic field inside the heart, suspended like a magnetically levitated train above its track. The control of the disc, which is doing all the work of moving the blood, has to be immediate and precise. The patient with the heart is hopefully well enough to be physically active, and that's a challenge; the disc is

---

* Many online sources claim Clark was the first—he was not. Karp was. By splitting hairs, you could argue Clark qualifies because the intent was for his artificial heart to be permanent.

unattached and can't be allowed to move out of position and contact the inside wall of the heart itself. So its position is always under control, with small bursts of magnetism correcting for any sudden movements. There is no pumping, just a smooth flow of blood through the heart to the lungs to pick up oxygen or to the body to distribute it.[5]

Where are heart transplants and artificial hearts headed? From the successful recipient's point of view, transplants are a godsend, on average giving them an additional decade of life. But given the huge amount of expertise, money, and technology that's been poured into both, neither has had a revolutionary impact on the treatment of heart disease. Transplants have advanced further, although progress continues to be slow because of cost and the lack of available hearts. (It's notable that in Germany, a country fully set up to deliver them, transplants are on the decline. Excessive paperwork has been identified as one of the main reasons.) Neither is the artificial heart a spectacular achievement—at least so far.

It's worth noting that heart transplants are in their sixth decade of development; artificial hearts are about to enter their fifth. *Newsweek*'s claim that the heart transplant was kicking off an era as important as the "age of the atom" is taking its sweet time to materialize. Will either the heart transplant or the artificial heart eventually prevail? Or could it be something else? We still don't know. Heart transplants might be given a boost by a new technique that allows a donor heart that has stopped beating to be revived and implanted in another patient. It's also been suggested that there is a viable third alternative to transplants: extend the use of ventricular assist devices to keeping the heart going while it receives treatments (stem cell implants, drugs to rebuild heart muscle, or even gene therapy) to recover from whatever is causing it to fail, then remove the device. If that were possible, the push for a permanent replacement might lose momentum. (However, in January 2022, surgeons at the University of Maryland Medical Center pulled off a medical first by transplanting a pig's heart into a human.[6] The heart had been substantially genetically altered to reduce the chance of rejection. If xenotransplantation—implanting an organ from one animal species into another—becomes viable it could be a partial solution to the shortage of hearts for transplant.)*

Prostheses

Prosthetic limbs are a different story. Yes, they have a much longer history—though not much progress to show for it—but in the last few years a combination of technology and biology has revolutionized prosthetics, both for upper and lower limbs.

Reach for the nearest object that you can easily pick up, and as you do, pay close attention to the feelings in your hand and especially your fingers. If you had your eyes closed when you picked up the object, would you have been able to identify it? How? Likely by shape, texture, temperature, or weight, but more likely by all four and their interaction. How did your hand have to shape itself to secure its grip? The amount of pressure you exert on the object to ensure it stays in your grasp has to be moderated by its fragility—if it's an egg, you don't want to crush the shell, but neither do you want to drop it. The neural circuitry that enables this simple act, one that is repeated countless times every day, is profoundly complicated. Sensory nerves in the hand, especially the tips of the fingers, record the force and "feel" of the contact between fingers and object. That information is relayed up the arm to the brain, where the object is identified, aided by memory of similar objects, and the next sequence of movements is determined. While that's happening, the position of your hand in space and its grip on the object are still being monitored.

Then a wholly different set of neural signals travels down the arm to the hand to execute whatever that sequence is to be. Imagine then transferring the egg to both hands and cracking it! Performing exactly the

---

* Nothing in this field is uncomplicated. Although none of the original accounts of this event mentioned it, it turned out that the heart recipient had been jailed years earlier for stabbing and paralyzing a man. Then the recipient died, and on autopsy it was discovered that the pig heart was contaminated with a pig virus!

same task with gloves on would be a completely different, more difficult, and not nearly as vivid a sensory experience.*

Now imagine doing all that with a prosthetic hand, or even a prosthetic arm. If there is no nerve supply to the artificial hand, there is none of the information necessary to calibrate movements and pressure, no feedback about texture or temperature. This is the world in which amputees have had to live. But amazing leaps forward are happening in labs around the world. Diverse research and clinical trials illustrate the dramatic advances in prosthetics.

Some people are born with foreshortened arms, while others lose part or an entire limb as the result of infection or accidents, especially with fireworks, electrical wiring, farm machinery, and explosions in conflict. The challenge with, say, an amputation between the wrist and the elbow is to create a prosthetic hand and lower arm that the wearer can control precisely. This necessitates a hand with fingers that can move independently, sensors connected to the nerves still active in the stump that can detect and convey neural messages quickly and efficiently to the wrist and fingers—those messages originate in the brain as the desire to pick up a pencil or turn a doorknob. To say it's a complex challenge is a gross underestimation: one crucial step is to translate electrical signals from muscles and nerves to digital signals to run the motors that control the movements of the hands and fingers. A computer, a power supply, secure connections from nerves to wires, all ideally self-contained—no external wires or power pack worn on a belt.

Some of the most striking recent advances have been giving the amputee a feeling of what the prosthetic hand is doing. At the level of neurons, the act of picking up an object is far from simple: different kinds of specialized neurons feed information independently to the brain about how they're being stimulated and that mix of sensation is

---

* That connection between hand and brain is an ancient one: the making and use of stone tools was one of the key events that allowed us to diverge from other primates. Recent experiments with university students have illustrated the intense brain activity associated with the movements and feel of shaping such tools.

integrated to provide you with the most complete picture possible of what you're touching. It is not a smooth stream of signals, either: there are sudden bursts of activity, both when you initially make contact and when you let go. That sudden onset and offset are critical to a realistic sense of touch and some prosthetic hands now do this. When that sensory feedback is available to the user, even though the sensations being relayed back to the brain might not be exactly the same as natural touch, they still provide enough information to allow the user to pick up an egg without cracking it, handle a grape without crushing it, and most important, be at ease with the artificial sensations. In the end—and there are numerous issues still either to be solved or refined—the goal is to have the prosthetic wearer use the arm/hand combination without thinking about it, like sitting and chatting with friends over coffee and raising a coffee cup, precisely and carefully, to your lips. While you talk.

There are numerous YouTube videos of just such actions. The prosthetic's movements are consistently smooth and natural. There is nothing robotic about it. It's only when you see an arm, by itself on a table, moving from side to side, making a fist and unclenching it, still controlled by its user, who's sitting on a chair nearby, that you realize this technology is truly bionic.[7]

Sadly, amputations of the leg are not uncommon, either, and sensory feedback from the prosthetic is just as important as it is for arms and hands. As you sit and read this, the only way you know the position of your foot—without looking—is by your brain getting information about the foot's location from signals that tell it which muscles are contracting and which are relaxing. Point your toes and your calf muscle contracts; extend your heel and muscles in the front of your lower leg contract and the calf muscle relaxes. The information about the speed, length of muscle, and force is relayed to our brain by nerves recording that muscle activity.

You may have temporarily experienced the loss of these sensations: sometimes if you're sitting cross-legged too long the circulation in your leg gets cut off, your foot is "asleep," and it's practically impossible to walk. You can't actually feel your foot until it "wakes up," a process that can be hard to endure sometimes, because you're lacking that same sort of critical feedback.

In the case of a lower leg amputation, improvements in the prosthesis can help, but improvements can be made in the amputation itself. The prosthetics group at the MIT Media Lab, led by American rock climber, double amputee, and biophysicist Hugh Herr, has taken a close look at amputation surgery itself and devised a technique to make sure that pairs of muscles in the stump are connected (most such pairings are usually severed during the amputation). These pairings act like the leg muscles that move your foot: one muscle contracts and its neuronal activity (translated by computers on the prosthetic limb) tells the ankle to extend, while simultaneously the companion muscle relaxes and *that* information is conveyed to the brain.

Their reciprocal movements translate the physical actions of a prosthetic into neural signals that reach the brain and give the user critical information about the position of the transplant.

The Ultimate Prosthesis

Hugh Herr himself lost both lower legs to frostbite after a climbing accident when he was seventeen. He has been involved in developing specialty prostheses for the lower leg and especially for the foot for people who, post-amputation, want to return to the life they previously led. One of his TED Talks refers to two such people: his longtime climbing pal Jim Ewing and ballroom dancer Adrianne Haslet-Davis. What he and his team have done for them is amazing. Ewing went back to high-risk climbing, taking advantage of a specially shaped prosthetic foot, while Haslet-Davis, a ballroom dancer who lost her lower left leg in the 2013 terrorist bombing of the Boston Marathon, danced again.[8]

It's heartening, thrilling even, to see Ewing climbing and Haslet-Davis dancing, but Herr, in praising the work of his team, made clear the enormity of the workload of customizing such prostheses. For Haslet-Davis, his team included a prosthetist, a roboticist, a machine learning specialist, and a biomechanics expert, who combined took *two hundred days* studying the movements of ballroom dancing, the forces exerted by the legs and feet, then importing that information into her prosthetic. That amount of expertise devoted to a single limb says that it can be done, but it also says that the vast majority of people will not be the beneficiary of such advanced prosthetic technology. Unfortunately, all advanced prosthetics are still a long way from being in general use.

Which brings us to the future of these technologies. Their purposes are different: the patient eligible for a heart (transplant or artificial) would not live long without it, while of course the amputee could just continue life as it is. In terms of restoring a good life, however, they are equivalent. But artificial hearts seem to be treading water. And though the most advanced prosthetics are still superexpensive and confined to the lab, there is an impetus behind them, and their promise is embodied in the technology: right now the best prostheses might have a handful of possible movements and sensations, but the electronics available can handle hundreds. (Of course, the living, biological version is capable of thousands.) You don't need a wild imagination to see them becoming cheaper, manufactured commercially, and beginning to have a huge impact on those who have lost a limb, whether by accident or genetics.

Both these technologies seek to replace what has been lost. And while I don't anticipate a proliferation of artificial hearts on the market, I do see prosthetics becoming more sophisticated, more capable, and more common. But a look further into the possible future suggests that technology might not be the only route to repairing people who have lost a limb.

We Could Aspire to Axolotlhood

Very few of us envy salamanders, but we should. They are the only vertebrates that can routinely grow back a limb. One of the family, the Mexican axolotl, can also regrow its brain, heart, lungs, spinal cord, and more. More remarkable is that as it regrows lost brain tissue, it reestablishes the broken neural connections.[9] One notable thing about the axolotl compared to its salamander brethren—it never really grows up. It doesn't lose its tail fin or gills as other salamanders do as they become adults. Whether this retained youth has anything to do with their ability to regrow a leg, who knows? There's no direct evidence, but the younger an animal (and this includes humans), the better the chances of regeneration. Children three or four years old who get a finger caught in a door and lose the tip of it will regrow the finger as long as some of the nailbed is still there. Those cases surely suggest that at least for a time in our lives we have the genetic programming to replace the tip of a severed finger, but only early in life and only the tip. An entire finger has never grown back. But fingers are child's play for the axolotl.

The regeneration of a limb in an axolotl likely requires neighboring cells to reprogram themselves and also stem cells to develop directly into nerve, muscle, skin cells—whatever is needed for the rebuilding, although it sure doesn't need many.*

The timing and logistics of axolotl regeneration is amazing but not completely alien—it resembles the unfolding of the original embryo. But at the point when an axolotl engages the regeneration program and starts to build a limb, a mouse develops a scar. Once scarring takes place, regeneration stops. Blood cells called macrophages participate in the regenerative process in axolotls but pursue the scarring approach in mice—somehow the macrophages in the two species respond differently to the signals of amputation's acute threat to life. The identification of the protein molecules that deliver that signal can't be far away, especially since the giant genome of the axolotl has now been sequenced and early observations suggest there are genes active at the amputation site that are not found in mammals.

A spectacular recent experiment demonstrated the regeneration of an amputated leg in *Xenopus laevis*, the African clawed frog. These animals are better stand-ins for humans than axolotls: they don't have a huge

---

* The flatworms called Planaria hold the regeneration record. They don't have legs to regrow, but their entire body can be cut in half, then in half again, and again, until each piece is no more than one-tenth of a cubic millimeter (about the size of the period at the end of this sentence). This for a worm that, when intact, is about 10–12 millimeters or about 0.4–0.5 inches long. The tiniest fragment of a millimeter of worm, containing maybe ten thousand cells total, can regenerate the whole animal.

supply of stem cells available to create new muscle cells, bones, and nervous tissue, and their ability to regenerate declines rapidly as they age, just as humans quickly lose the ability to regrow the tip of a finger. The frogs can regenerate their tails as tadpoles, but not legs as adults. At least, regeneration of a limb had never been seen before. But now it has.[10]

In this latest successful attempt, the stump of the frog's leg received a dose of a set of drugs known to be crucial for regeneration, delivered by what the experimenters called a "wearable bioreactor," a silicone pad containing a silk-based gel into which the five drugs were implanted.[11]

Maybe the most amazing thing about this experiment was that the pad was left on for only twenty-four hours, but the regeneration in the drug-treated frogs continued for months. After eighteen months they were significantly different from the controls, having grown new bone, muscles, and nerves and created a good facsimile of a normal hind leg, capable of both movement and sensation. It wasn't perfect, but good enough to suggest that fine-tuning the process could eventually enable full, natural regrowth. In a sense, what the experimenters had managed to do, in a mere twenty-four hours, was reestablish embryonic conditions in the frog's stump, which then persisted for months, accomplishing what embryos do—growth and development.

The scientists want to move on to mammals next, and they will likely be an even greater challenge. But axolotls aside, these experiments brought the distant dream of regrowing parts of the body a little closer. It will require a major rewriting of biology. At the moment, the research isn't far enough along to anticipate the possible roadblocks ahead, let alone surmount them.

Most of what I've described are the efforts to restore lost function. But those boundaries will inevitably be pushed to see replacement superseded by enhancement. Yes, there are precedents, like Oscar Pistorius, the double amputee sprinter (and convicted murderer), whose ski-like blades made it possible for him to run a sub-11-second 100 meters and compete in both the Olympic and Paralympic Games. But his prosthetics are straightforward and already much copied. The question is, what else might we see in the future? There are already glimpses of an answer.

CHAPTER 3

Cyborgs

Most of what I discussed in the previous chapter would fall into the category of "replacement" or "restoration"—giving back something that has been lost. Most existing technologies fall short of completely restoring lost abilities: hearing aids don't restore the hearing you once had; a prosthetic limb fails to give you the mobility you had as an unimpaired teen. However, as these replacement technologies improve, it's easy to imagine that at some point "enhancement" may be possible—a human with abilities greater than he or she ever possessed. It's a popular science fiction theme, and it is edging ever closer to reality.

"Gentlemen, we can rebuild him; we have the technology." That was the crucial sentence in the introduction to the 1970s hit TV show *The Six Million Dollar Man*.[1]

Colonel Steve Austin, played by Lee Majors, survived the crash of his experimental aircraft but lost both legs, his right arm, and his left eye. All were replaced by so-called bionic implants. He was, as the narrator claimed, "the world's first bionic man."

His limb and eye replacements did more than simply restore Austin's function. They furnished him with superhero abilities: his replacement eye had a zoom lens, his legs gave him Usain Bolt+ speed of at least 100 kilometers per hour (65 miles per hour), and his arm had the power of a bulldozer. Unfortunately, nowhere was it explained how the rest of his body, unenhanced by technology, could support these feats of speed and strength. (*The Bionic Woman*, starring Lindsay Wagner as Jaime Sommers, was similarly enhanced with a bionic right ear, right arm, and two incredibly powerful legs.) This was TV, not reality, and while the TV world has elaborated on this theme over and over again, nothing in the real world has come close.

As I wrote in the previous chapter, replacement arms and legs—especially the hands and feet—are improving on a very steep curve, and twenty years from now such limbs aren't going to look, or act, much like what is available today. But are they likely to develop along Steve Austin lines?

Build on the Human

To an insect, an exoskeleton is the shell in which its body is encased. It's their protective carapace (insects that molt and temporarily lose their exoskeleton are easy prey) and a necessary element for support and movement. Exoskeletons for humans are similar: protective in the work environment because they lessen back strain for those who lift heavy objects and supportive because they do much of the muscle work in lifting those objects. It's envisioned—and already happening—that they will have wide-ranging application, including recreational activities like running, but also rehabilitation medicine, warehouse labor, and the military.

There is overlap among these different applications, although the detailed designs for each are very different. Relatively simple exoskeletons limited to the ankle and foot have shown that providing additional power this way could allow runners to move 10 percent faster without any effort on their part. Doubtless serious amateur athletes will take to these exoskeletons once they're commercialized, as many have to electric bikes. People with spinal cord injuries recover some of their lost mobility while encased in an exoskeleton that provides them support and the power to move. Industrial exoskeletons allow a person to lift much heavier weights without risking lower back injury, while military exo-skeletons are designed for soldiers with heavy gear to move for longer periods with less fatigue.

Exoskeletons like these are just now being rolled out, some closer to market than others, but there's a crucial difference among them. While exoskeletons for spinal injuries or other movement impairments are designed to regain function, the others are among the first intended to *enhance* function. Replacement versus enhancement is a major test for the integration of technology with biology.

Many futurists, technophiles, and science fiction writers have antici-pated these times, when humans and machines merge, but who has the most accurate take on what might happen? The Massachusetts Institute of Technology's Hugh Herr, whose work on prosthetic limbs I described in the last chapter, has overseen some of the most rapid advances in prosthetics and he has been powerfully moved by that experience: "I believe that the reach of neuro-embodied design will extend far beyond limb replacement and carry humanity into realms that will fundamen-tally redefine human potential. In this twenty-first century, designers will extend the nervous system into powerfully strong exoskeletons that humans can control and feel with their minds. Muscles within the body can be reconfigured for the control of powerful motors and to feel and sense exoskeletal movements, augmenting humans' strength, jumping height, and running speed. In this twenty-first century, I believe humans will become superheroes."[2]

Bold. Herr also quotes Leonardo da Vinci's "eyes turned skyward," and he sure is doing that, looking way ahead. When he says "exoskeleton" he's not talking about those I've already mentioned. His exoskeletons are ones that "humans can control and feel with their minds . . . muscles can be reconfigured for the control of powerful motors." If he's right, that would put us in the same ballpark as the bionic man and confirm Herr as a visionary.

I think Herr could be right. He knows the technology; whether his most daring predictions will come true, who knows? He is a powerful advocate—that matters—and his main idea, that there will be a merging of the human and the synthetic, is already in a form of existence with prosthetic limbs. It's easy to envision a day when you won't be able to tell those who have prosthetics from those who don't. Or, they will be completely personalized—no two alike.

CYBernetic ORGanism

"For the exogenously extended organizational complex functioning as an integrated homeostatic system unconsciously, we propose the term 'Cyborg,'" wrote Austrian-born scientist Manfred Clynes and American psychiatrist Nathan Kline in the journal *Astronautics* in 1960.[3] Thank God they did. Beyond the polysyllables, what did they mean? They envisioned humans deliberately adding foreign components to their bodies, becoming partly synthetic, partly biological—cyborgs. Or as they put it, "Altering man's bodily functions to meet the requirements of extraterrestrial environments."

The *cy* is for cybernetics, explained by the subtitle of American mathematician and philosopher Norbert Wiener's book *Cybernetics, Or, Control and Communication in the Animal and the Machine*. Considered the founder of cybernetics, Wiener envisioned the world as being built on and by networks communicating by feedback. *Cyborg* is short for "cybernetic organism." In their article, Clynes and Kline were looking ahead to a time when humans would be living in space, and they wondered how human physiology might be altered to adapt to such an alien environment. They walked through a variety of spaceflight challenges—oxygen supply, muscle maintenance, mental health, radiation—all with an eye to making things better by melding manufactured elements with the biological. It could be something relatively straightforward, like an implanted device to time the release of a chemical into the bloodstream, but Clynes and Kline imagined other possibilities, such as rebuilding the digestive system so that urine is treated and then released back into the veins, never excreted, while the bowel is sterilized and pretty much shut down. Nothing excreted there, either. Very few of their speculations are attracting any attention these days, but *cyborg*! The term is as common in pop culture as it is in science: the Terminator, Darth Vader, Robocop, the Daleks, the Replicants, the Borg. My Steve Austin/Jaime Sommers examples earlier are pretty dated, but they had an impact: most of us know of the Six Million Dollar Man and the Bionic Woman.

Those are the television and cinematic versions. In the real world, things are different. Here, again, we run into the distinction between repair/reconstruct and enhance. The elaborate prosthetics of the last chapter are almost always the former (without ruling out the possibility that they could be tailored to improve performance). But a pacemaker isn't designed to allow you to break track records. There is a fuzzy line here: Peter Scott-Morgan, the English roboticist who died of amyotrophic lateral sclerosis, or Lou Gehrig's disease, was an example. He progressively added technologies in an attempt to replace functions lost as his motor neurons died. He designed a self-driving robotic exoskeleton, which would have made him more powerful than he was. His goal was to live a long, fulfilling life, not to become ultrahuman. But it's the enhancement piece that sparks the imagination and tempts futurists. Cyborgs as *more* than human.

They Walk Among Us

Are there such cyborgs now? There are, although while a few made headlines at first, they're mostly flying under the radar now. One who has actually taken the technology somewhere is Neil Harbisson. In 2004, Harbisson, a color-blind artist, had an antenna surgically implanted in his skull. (He was unable to get ethical approval for the surgery so he had to find a surgeon who would do it undercover.)[4] The antenna converts the varied wavelengths of colors (even extending to the normally invisible infrared part of the spectrum) into vibrations that Harbisson can feel through his skull. The antenna also connects to the internet. Harbisson is by now expert in differentiating subtle shades of color by their vibrational pattern and has initiated several exhibits based on his unique talent. Some of these exhibits are for an audience of one—himself. But he's experiencing cyborg life, as he told me when he appeared on *Daily*

Planet: "After a few months of wearing it I began to feel that the eye was part of my body—it just became an extra sense."[5]

One who's closely watched Harbisson's experiment is English engineer Kevin Warwick. Warwick himself is a cyborg. He started modestly enough by having an RFID chip implanted in his forearm in 1998, thus enabling him to unlock doors that you'd normally open with your employee card. He wasn't finished. His next step into cyborgism was to have an array of electrodes implanted in the medial nerve in his left forearm at the Department of Neurosurgery and Neurosciences at John Radcliffe Hospital, in Oxford, England. The array was just below the wrist. A wire extending from the array tunneled under the skin, emerging from a second incision about halfway up Warwick's forearm, connecting to a terminal that remained outside his body. The terminal in turn was either hardwired to a computer or connected wirelessly. Warwick wore the whole setup for more than three months.

He and his team conducted several experiments during that time, two of which Warwick thinks are most significant. In one, Warwick was at Columbia University in New York connected via the internet to a robot hand in Reading, England. Signals from his implanted array of electrodes were transmitted via internet to the robot hand. When Warwick opened and closed his fist, the robot hand, after a short delay, did the same. It was a dramatic moment. But then Warwick went one step further and persuaded his wife, Irena, to have an implant in her arm so the two could communicate electronically (more powerful than tattooing your partner's name on your arm!). When Irena closed her hand, Kevin felt a pulse, and vice versa.

These experiments might seem pretty simple, but it's not so much what they've accomplished as what they promise.[6] Warwick has shown that an implantation of a set of electrodes allows meaningful information to be recorded and transmitted either to a robotic hand (relatively straightforward) or even another person (much less straightforward). Warwick sees this early work as demonstrating the connection of one human's nervous system to another's and makes the leap to connecting one brain to another brain directly. That's obviously much more challenging, invasive, and perhaps threatening than inserting electrodes into the medial nerve of the forearm. But not only is such an idea on the minds of many, connecting the brain has already been done.*

Extracting useful information from the brain using sets of electrodes seems strange to me. We're told the complexity of the brain is somehow greater than the entire Milky Way galaxy, but then we hear of people like Harbisson and Warwick who have used relatively straightforward technology to augment the natural abilities of their brains. To me, the technology seems primitive, the connectivity bare-bones compared to the complexity of neurons/synapses, but a quick look at three much-quoted experiments will give you a feel for just how we're doing in establishing communication with the brain.

There Are Experiments!

One experiment used a technology called BrainNet. BrainNet's inventors call it "the first multi-person non-invasive direct brain-to-brain interface for collaborative problem solving."[7] BrainNet allows three people to play a *Tetris*-like game simultaneously. Two players "advised" the third, but no one spoke, no one saw anybody; each player simply stared at a screen. The advisers had targets on their screens where they could either vote "Yes, rotate the piece" or "No, don't rotate the piece" (before it's dropped into place). The brain activity associated with those thoughts was detected by electrodes embedded in a cap the players wore, transmitted to the player by transcranial magnetic stimulation (a way of uploading information to a brain), set up so that a yes would create an internal flash of light (like "seeing stars") and a no would not. Each turn even allowed a moment for rethinking the planned move.

Teams were better than 80 percent in landing the piece properly,

---

* Then there's Chris Dancy, "the most connected man on earth." His website claims that he's used hundreds of sensors and devices to "see the connections of otherwise invisible data, resulting in dramatic upgrades to his health, productivity, and quality of life." Nathan Copeland, while he doesn't advertise his connectedness with quite the same fervor, has had a brain implant since 2015.

even though the "player" had never seen its final resting place and was just going on the advice of the other two players, delivered electronically. You might see this experiment as an unimpressive combination of complex technology being used to accomplish a tiny task, even a little Rube Goldberg–ish, but it was brain-to-brain and noninvasive. The latter is impressive when you see how many other designs require surgical implantation.

Of course, our most pressing need is to serve those who can't readily communicate, especially people who are paralyzed and have lost the ability to speak, but whose brains are absolutely normal. Uninjured people can type at 12 to 19 words per minute; touch typists 40 to 60 words per minute; and speaking lays down 90 to 170 words per minute. In several recent studies, people who have arrays of electrodes implanted in their brains manage about 6 to 8 words per minute by choosing letters displayed on a screen. Not bad, but obviously not where they'd want it to be.

This performance was exceeded recently by a man who created an alphabet of his own handwritten letters.[8] He had an electrode array implanted in that part of the brain that is primarily responsible for hand movements, and then he imagined how he'd write each letter. Once those patterns of neural activity for each letter had been interpreted and recorded by the computer, he could then think of the letters to write at a rate of 18 words per minute with 94 percent accuracy—better than I can do on my phone. There's something about handwriting, which is highly variable and almost never has straight lines, that is interpreted more easily by software than the straight lines traced by a cursor on a screen.

Speaking of interpreting one's thoughts, here's a final example of an experiment that demonstrates that what you're thinking about can be decoded directly rather than being printed out letter by letter. Decoding is necessary because all brain activity is electrochemical, so thinking of, say, today's breakfast, will generate electrochemistry in certain places in your brain, and the objects associated with those patterns might be eggs, toast, coffee, granola, whatever. But the patterns don't in any way *resemble* the things your mind is envisioning.

Psychologist Jack Gallant's lab at the University of California, Berkeley, has demonstrated that magnetic resonance imaging (MRI) of humans' brains as they watch an assortment of movie clips can be decoded to identify the objects in those clips the brain is paying attention to or thinking about.[9] The experimenters set the stage by cataloging the objects or actions present in every second of every scene in a selection of movie clips. In the first round of the study, a group of volunteers watched the clips while their brain activity was visualized using MRI. While MRI is not very good at timing neural activity, it can *locate* such activity precisely. In this experiment the MRI demonstrated that there were many locations in the brain associated with specific objects or actions. Verbs like *drag, move,* and *breathe*, nouns like *city, flower, forest*, and hundreds more triggered responses in specific places in the brain.

With that map of where actions and objects reside in the brain, the next step was straightforward. The volunteers watched a never-before-seen trailer of clips. Again, the MRI recorded their brain activity. But this time the scientists, who were not familiar with the film, tried to identify some of the content of the film just by scanning the MRI imagery and looking for telltale peaks of activity in familiar brain locations. Intense activity in this location at the three-minute mark? That likely means there's a car on-screen.

These three studies illustrate different ways of transferring thoughts to and from the brain.* In 2022, a man who was completely paralyzed, unable even to move his eyes to communicate, was able to use implanted electrodes in his brain to assemble letters to utter complete sentences, including the memorable "Boys, it works so effortlessly." He is the first completely "locked-in" person to have been able to do this. But even this dramatic advance illustrates that it's still early days: for reasons that aren't

---

* It's worth remembering professor of neurophysiology José Delgado here. In 1965(!), in a bullring in Spain, he stopped a charging bull in its tracks. The bull had an electrode implanted in its motor cortex, that part of the brain directing movement. When Delgado pressed a button on a transmitter connected to that electrode, the bull immediately stopped. The bull was charging at Delgado at the time.

yet clear, his ability to express himself eventually regressed to simple yes and no responses.

So there's still much work to be done and the technology that eventually becomes capable of seamlessly transferring thoughts to or from a brain will be like none of the above, but you can't deny it's coming. In fact, at the time of writing a research paper (not yet peer-reviewed) a group at Stanford University claimed a record performance of a brain interface by a man with ALS: 62 words a minute with an acceptably low error rate. It's hard to doubt the authors' claim that "[t]hese results show a feasible path forward for using intracortical speech BCIs [brain-computer interfaces] to restore rapid communication to people with paralysis who can no longer speak."[10] The question is, do we have any idea what that human-brain-to-machine or human-brain-to-prosthetic or human-brain-to-human-brain communication will look like?

The Ultimate Connection

Both "neural dust" and "neural lace" are hot topics in the brain-computer-interface world.

Neural dust, as the name suggests, takes the concept of implanted electrodes to an ultra-small dimension. Neural dust particles contain crystals that respond to ultrasound—in fact, they can be powered by pulses of ultrasound, eliminating the need for onboard batteries. That's a huge step forward because batteries significantly increase the size and decrease the longevity of implanted sensors. Just as important is the fact that these particles, which are the size of the tiniest grain of rice, can transmit information about their surroundings back to an external receiver by the same ultrasonic waves. Ultrasound beats electromagnetic waves in this case because it penetrates tissue much more easily and is much less likely to overheat surrounding tissue. Despite their diminutive size, each "mote" of neural dust contains the crystal, a transistor, and two electrodes. They communicate with a device called an "interrogator" located on the brain's surface.

There are still issues to overcome with neural dust—like ensuring that their presence doesn't evoke unwanted reactions to the tissue in which they're embedded and that over time they remain intact and undamaged—if they are to have a role in the future.

The also cleverly named neural lace takes the existing technology—an electrode-studded EEG helmet or cap pulled tight over the scalp—and transforms it by shrinkage. Imagine a microscopic fishnet or mesh, an electrode at each node rolled out over the surface of the brain. The mesh can be injected with a syringe and will take up residence in the tissues without being rejected. Because it's an open mesh, cells can penetrate and grow inside it. At this point the promise of neural lace seems to be an adaptable vehicle for a variety of uses, but the brain is an obvious target: implanting hundreds or even thousands of recording electrodes would change the brain-computer-interface game.

Even that notorious game changer Elon Musk has taken note of neural lace—I think.[11] In an interview in 2016, he repeated his contention that as artificial intelligence becomes more and more intelligent, we will need some way of interacting with it to protect ourselves from it (much more on this in Chapter 15). In Musk's view, just as the cerebral cortex of the brain sits atop areas like the limbic system (a shout-out to the old idea of the reptilian brain), we should invent a new digital layer to be draped over the cortex to interact directly with artificial intelligence (AI). In the same interview he referred more than once to neural lace. But the products so far revealed by his company Neuralink are not neural lace at all, but coin-sized brain implants with a reasonably large number (1,024) of fine wires extending electrodes through the brain. By the time you read this, one or more of these coin-sized implants might have already been implanted in human brains.* Neuralink has done it with a pig named Gertrude, and while she remained healthy she wasn't given a task to perform with her mind. Pager, a nine-year-old macaque monkey with

---

* At best they'll get the silver medal. A company called Synchron had already implanted their technology in two patients by the summer of 2022. Neuralink had announced plans to start clinical trials in the fall of 2023, but approval was delayed by a US federal probe into alleged violations by Neuralink of the Animal Welfare Act.

a Neuralink disc in his brain, learned to play *Pong* using a joystick, then continued to play after the joystick was removed, the activity of his brain controlling the game directly. In late 2021, Musk claimed that brain-to-computer interfaces using Neuralink could exceed even the 90 to 170 words per minute of conversation, and more controversially stated, "I think we have a chance with Neuralink to restore full-body functionality to someone with a spinal cord injury."[12] That, of course, would be remarkable. Neuroscientists are generally skeptical of that claim, as they are of Musk's assertion that Neuralink will be able to restore vision to the blind.

It may have struck you already that implanting a device in the brain is a dramatic medical step, yet some versions of the technology are already practically routine. More than 100,000 people with Parkinson's disease have had electrodes implanted in specific target parts of their brain to facilitate movement. This procedure, called deep brain stimulation, is also being tried on those with obsessive-compulsive disorder, tremor, depression, epilepsy, and even chronic pain.

But . . .

Even though deep brain stimulation is already well established (although brain implants are not yet what you could call routine), there are concerns beyond the risk of surgery that will only become heightened as different types of brain implants are attempted. One is ethical: after implantation, the patient may feel different, or not themselves, or that they no longer control their moods and actions as they once did. Such reactions have already been reported, and you might imagine that increasing the numbers of electrodes and the locations in the brain where they're placed will only increase the number of patients who have this experience. An even more dystopian view has been expressed by Susan Schneider, a philosopher at Florida Atlantic University, who argues that having a chip in your head opens up the possibility—in the absence of adequate regulations—that a company could monitor your brain activity, or even collect it and sell it.[13]

The other concern is "brain jacking." In deep brain stimulation, the electrodes in the brain are controlled by an impulse generator implanted under the skin in the chest. If the impulse generator's software can be hacked, then the hacker could do real damage, either by withholding the necessary stimulation or amplifying it beyond safe levels. Hacking a heart defibrillator might be easier and more effective.* As always, the deployment of new technologies is only partly dependent on the technology itself. Social, ethical, political, and financial issues are always at play.

At this point in the invention and development of these brain-computer-interface technologies, it's really impossible to say which design(s) will flourish. No medical device aimed at communicating with AI will be approved anytime soon, but new ones that allow people to communicate likely will. After all, the first brain implants are already in place.

Who better to speculate how that first step will unfold than Hugh Herr, the man who has overseen dramatic progress in the development of artificial limbs? When Herr looks many decades ahead, he wonders if the changes we will see are almost beyond comprehension today: "Humans may also extend their bodies into non-anthropomorphic structures, such as wings. Controlling and feeling each wing movement within the nervous system. In the twilight years of this century, I believe humans will be unrecognizable in morphology and dynamics from what we are today. Humanity will take flight and soar."[14]

---

* In the TV series *Homeland* this is exactly how the US vice president was killed. Years before that, when Dick Cheney was the actual vice president, he had the wireless feature of his implanted defibrillator inactivated to prevent a remote signal being sent to the device to trigger cardiac arrest.

CHAPTER 4

It's in Our DNA

Sometimes it's called "synthetic biology," but it seems bigger than that, a radical combination of biology and engineering, an effort not to just read and understand the book of life but to edit and rewrite it. DNA as a workbench tool. Where we stand today, DNA is positioned to create absolutely revolutionary advances in many different branches of science, certainly most that I'm touching on in this book. Imagine being able to write a genetic sequence on a computer, then turn around and have that sequence realized in DNA and inserted into the genome of a microorganism to produce the product you want. Then after a few hundred cell divisions, the recipient bacterium and a few billion of its colleagues have become a factory. That's where we are at the moment. Aging? There will be synthetic biological approaches to slow that process down. Alzheimer's disease, cancer, invasive species, climate change? There's a very long list of human challenges that could be addressed by combining the best of technology today, especially artificial intelligence, and the evolving ability to manipulate DNA. Of course, at every step of the way, given the centrality of DNA in life, there are ethical and equity concerns. But even given the likely stop-start nature of such revolutionary possibilities, there seems no doubt that we're entering a time when what was thought impossible just might become possible and what hadn't even been imagined is now on everyone's mind.

By the time you read this it will have been seventy years since James Watson and Francis Crick published their historic paper in *Nature* introducing the famous double helical structure for DNA. The discovery, an intellectual tour de force, was tarnished by the fact that the two used data from British chemist Rosalind Franklin without her knowledge. While Franklin did publish that data in a companion article in the same issue of *Nature*, she didn't receive the Nobel Prize that Watson and Crick did (although, by Nobel rules, she couldn't have, because she died before the prize was awarded in 1962).

It might surprise you that the discovery of DNA was somewhat muted at the time. I've talked to scientists who attended biology-related meetings shortly afterward and claimed there was no real buzz about DNA at all. But today? DNA is solidly at the center of all the biological sciences, as it will be in the future.

It's the "genetic blueprint," although that bad analogy needs updating as the science advances. A blueprint is a two-dimensional plan. DNA requires a 3-D real-time animation. It is constantly creating new products—protein molecules the cell needs—by piecing together the protein building blocks called amino acids. Proteins can be large, complex molecules, but they are so precisely sequenced and assembled that they are obliged, chemically, to fold into the shape that tunes them perfectly for their role in the cell. They might catalyze chemical reactions; they might be a piece of the architecture of a larger complex; they could roam the circulatory system. The list is very long.

By the 1960s, scientists had a pretty detailed picture of how DNA works. It wasn't long before many influential scientists began to push the idea of sequencing the human genome, that is, the entirety of human DNA, every single identified gene. Without the complete genome, searching for genetic links to both health and disease is working in the dark. The Human Genome Project began in 1990; by 2000, the international team of geneticists and biochemists had released an early draft of the (nearly) complete set of human genes. Finally, in April 2022, a paper published in the journal *Science*[1] announced they now had a seamless chromosome-by-chromosome, tip-to-tip assembly of human DNA, excluding only the male Y chromosome because the cell system that generated the DNA didn't produce the male chromosome.*

The Human Genome Project labeled itself "one of the great feats of exploration in history . . . Rather than an outward exploration of the planet or the cosmos, it was an inward voyage of discovery."[2]

There's some truth to the discovery angle when you read in the *Science* paper that the main advance was the ability to sequence the five odd-shaped "acrocentric" chromosomes, whose two arms are joined very near the tips. The DNA in those very short arms just beyond the join had proved hard to sequence, but now it's done. Adding the DNA from the short arms of those five acrocentric chromosomes was equivalent to adding an entirely new full-size chromosome.

In one important sense the Human Genome Project was only a partial discovery, given that 70 percent of it was based on the DNA of one man from Buffalo, New York. There are efforts now to create a human "pangenome," which would incorporate the variety of "normal" genomes that exist in populations around the world.

You can now get your own genome sequenced by companies like 23andMe, and DNA has become part of the daily lexicon, often inappropriately so: "Playing good defense is in the DNA of our team." (No, it's not.) But consider this: the twenty thousand or so genes of the human genome represent a mere 1 percent of the genes we have in our bodies. One percent! The other 99 percent belong to what's called the second human genome, that of the "microbiome."

---

* The Human Genome Project hogged the headlines when it released its first results, but we shouldn't forget that a private company, Celera Corporation, led by Craig Venter, also sequenced the genome at the same time and did it for much less money.

Helpful Stowaways

The *microbiome* is comprised of the bacteria, fungi, and viruses that live on and inside us. Why should we count their genes as part of our genome? After all, these microscopic entities have their own life cycles, their own metabolisms, they can come and go, and most of them aren't what you'd call an integral part of our bodies. All of that is true, but it's also true that they are integral to our health.

The study of the human microbiome is charging ahead: the official Human Microbiome Project began in 2007. The challenge is substantial: our microbiome contains 10 trillion bacterial cells, slightly more cells than make up our bodies, representing a thousand species. As microbiome researcher John Cryan pointed out in a TED Talk: "When you go to the bathroom, and you shed some of these microbes, just think: you are becoming more human."[3] As Cryan implies, much of the focus these days is on the microbiome of the gut (promoted as it is for an almost endless variety of health benefits), especially the bacterial part of it, which incidentally weighs almost a kilogram, or nearly 2.2 pounds. The gut microbiome is the richest, but the skin and respiratory microbiomes, including not just their bacteria but fungi and viruses, too, conspire to make the Human Microbiome Project a massive undertaking.

Why the excitement? There is a vast number of studies and experiments that show that the bacterial community in our gut is responsible for much more than embarrassing eruptions of gas, although that is a downstream effect of their activity. Changes to the bacterial inhabitants of the microbiome after a prolonged exposure to antibiotics, for example, are implicated in a variety of gut conditions, like colitis. After all, that is their home territory and changes are often manifested locally. But the stunning thing about the gut microbiome is its potential to disturb at a distance—the prime example being something called the "gut-brain axis." It's estimated that of the communications between the brain and the gut, 90 percent travels from gut to brain.

Evidence that bacteria in your intestine can influence the brain comes from two kinds of observations: one, animal experiments (always accompanied by the disclaimer that they might translate imperfectly to humans); and two, human studies that are suggestive but have not yet been completely confirmed by experiments.

The evidence with animals, usually mice, is convincing. So-called germ-free mice, which are raised in sterile conditions and lack resident bacteria, exhibit abnormalities in almost every part of the brain, including neurotransmitters, their receptors, and synapses, the connections between neurons. This effect is, for mice anyway, intergenerational. At the University of California, Los Angeles, the American neuroscientist Helen Vuong conducted experiments showing that germ-free mother mice were carrying embryos whose brain development was abnormal, especially in areas concerned with sensory perception.[4] After birth, the young mice were unable to respond normally to heat, sound, or touch. Reconstituting a mother's depleted microbiome, even partially, restored the fetal mouse's normal brain development.

The exact mechanism underlying the gut-brain axis has not been determined, but it will almost certainly involve small molecules making their way to the brain after being generated by bacteria in the gut, or some sort of nervous tissue connection. Animal models of amyotrophic lateral sclerosis and circumstantial evidence of human microbiome disruptions accompanying that disease also demonstrate a likely gut bacterial connection. But finding that gut issues accompany these conditions—and others—doesn't necessarily establish that they *cause* them.

An Australian research team recently reviewed worldwide literature on the link between diet and depression and found that diets with multiple sources of polyphenols, like tea, coffee, nuts, soy, and legumes, were associated both with a lower risk of depression and a relief of symptoms.[5] The Mediterranean diet is loaded with polyphenols and has a reputation for being an exceptionally healthy diet.

Is this correlation or causation? The answer awaits clinical studies. But there are other studies, like one showing that depressed people lack two or three bacterial species in their gut that are common in people without depression.

Microbiomes, once established, tend to maintain themselves, but they can be altered, for better or worse. (We get our first blast of microbes from our mothers, and then augment it from our social contacts.) Diet is a powerful influence. Extended use of antibiotics, which of course target bacteria, can change the microbial landscape in the gut with serious consequences. But those sometimes profound changes can be reversed in unlikely ways, and sometimes such changes can be seen in the brain as well as the gut.

One study revealed that women who consumed fermented milk with probiotics demonstrated significant changes across a variety of neural networks in the brain. Deciphering what such networks do isn't easy, but both probiotics (foods containing bacteria for the gut) and prebiotics (foods containing nutrients that feed those bacteria already in the gut) are implicated in changing the brain. For instance, human breast milk is unique among mammals. It contains roughly two hundred different kinds of sugar molecules, whereas other mammals' breast milk contains fewer than fifty. But the twist is that a newborn cannot digest all these sugars; instead, some of the bacteria that are rapidly colonizing a newborn's gut metabolize them and gain a foothold in the rapidly growing microbiome. These sugars are not intended for the baby; they're intended for the microbiome, which in turn will help establish future health for the baby!

"Swallowing faeces is quite controversial for most people" is a quote I've always thought should kick off a paragraph.[6] It's attributed to gastroenterologist Christian Lodberg Hvas of Aarhus University in Denmark. He's referring to fecal transplants, a technique that is exactly what it suggests: transferring a healthy set of gut bacteria into the gut of someone who is suffering from extensive changes to the microbiome. It's not quite what it sounds like: a small dose of fecal bacteria can be delivered either directly to the intestine by a colonoscope or even to the stomach by swallowing an oral capsule. More important is the fact that these are usually highly effective. People suffering from colitis caused by the bacterium *C. difficile* are cured 90 percent of the time, a powerful demonstration of the positive health effects of the right microbiome. Some animal experiments suggest even more exciting possibilities—some age-related declines in old mice have reversed after they received fecal transplants from younger mice.

While we're familiar with sayings like "You are what you eat" and "It's a gut feeling," until a few years ago no one knew that both of these statements were literally true. So completing the Human Microbiome Project could have as much impact on human health as the Human Genome Project. Both are genetics on a grand scale.

Tools of the Trade

Reading the genetic message is one major breakthrough, but there are scientists who argue that the next big thing will be the *writing* of the genetic message—choosing the genes you want to put together, then assembling them. The concept is no different than picking words, one by one, out of a paragraph and assembling them in the order you want, either to sharpen the message or change it completely. Extreme precision is necessary to remove a single gene from a long stretch of DNA encompassing many genes, then inserting that gene somewhere else and have it work as you expect it to. That precision is provided by the technology called CRISPR.

CRISPR-Cas9 is the ten-year-old, Nobel Prize–winning set of tools that has revolutionized DNA research. The acronym stands for Clustered Regularly Interspaced Short Palindromic Repeats—a term so precisely descriptive as to make it extremely difficult to remember. It refers to short stretches of DNA repeated over and over in a bacterium's genome. Between the repeats are DNA sequences that have been highjacked from viruses that had previously attacked those bacteria. The bacteria apparently store these sequences in order to recognize those same viruses and deploy defenses against them. A pretty cool strategy.

In the case of a repeat viral invasion, the bacteria counterattack by using these sequences to guide a DNA-cutting enzyme to inactivate the genes of the attacking virus. In the more benign atmosphere of the lab, CRISPR can cut any sequence of DNA from any source with a precision that was impossible before. Cas9, or "CRISPR associated enzyme number 9,"* is the equivalent of the bacterium's DNA-cutting enzyme. CRISPR Cas9 can be used either to remove a gene (a stretch of DNA) from or add a gene to an existing DNA sequence. The technology has not yet been perfected, so caution surrounds its medical use. But in the lab CRISPR is a miracle.

The ability to edit DNA opens up as many ethical questions as it does opportunities. There is already an example: In 2018, Chinese scientist He Jiankui announced that he had genetically altered twin embryos for a Chinese couple. The father was HIV positive, and Jiankui used CRISPR to eliminate a gene in the twins known to facilitate the entry of HIV into susceptible cells. I don't know how he could have thought doing this would be uncontroversial; I doubt he anticipated a three-year jail sentence and a half-million-dollar fine for what the Chinese government called unethical behavior, including forging documents. The immediate international pushback against what the science community deemed the reckless use of a genetic technology was reassuring, although the incident also made the threat of rogue genetics a reality, unlike many other speculations about the future.[7]

Nature's Twists and Turns

While most of us hear about the genetics and the genomes of the new biology, the impact of DNA is indirect. As I said earlier, it's really all about the proteins. They represent the set of tools deployed by the genome to sustain life and they have a complexity that DNA doesn't have: they fold into complex 3-D shapes, the tiniest detail of which—an edge that juts out here or a cleft there—determines whether the protein works or not. The sequence of their components, the amino acids, determines the shape of the final molecule.

A single amino acid substitution in the long chain making up a protein can render the protein useless. Not because the sequence changed, but because the change in sequence changed the shape. Knowing exactly

---

* There are many others.

how the shape changed is important, but figuring out the exact folding of a protein or predicting it is extremely difficult. It was once calculated that the total number of folding possibilities for a medium-size protein of 100 amino acids would be 2^{100}, an obviously ridiculous number. There's no way nature can check out every possibility, given that at this moment thousands of such proteins are folding every instant in your body. There's usually one ideal shape—in a matter of nanoseconds, the protein will shake, shiver, twist, expand, contract, suddenly stop, then wiggle again before settling into its final form.

This is a field on the move. In July 2022, the AI group DeepMind made a sensational announcement about their program called AlphaFold, which can predict the final shape of a protein, not quite from scratch, but from even a small piece of the structure that's already known.[8] AlphaFold had determined the final shape of more than 200 million proteins, human and otherwise—which accounts for almost all the known proteins in nature.

AlphaFold does in minutes what it used to take months or even years to do, depending on how far back you want to go. So research will accelerate dramatically. For instance, AlphaFold has already made it possible for a team of parasitologists at Oxford University to set aside its ongoing efforts to determine the folding of a key protein on the malaria parasite—down to the location of each atom—and instead start with AlphaFold.[9] They're confident there are sites on that molecule that will provide new targets for antibodies to attack.

A diverse group of scientists at the Centre for Enzyme Innovation at the University of Portsmouth is using the technology to "deliver transformative enzyme-enabled solutions for the circular recycling of plastics."[10] They envision proteins, in the role of enzymes, digesting single-use plastics, the proceeds of which would be recycled or even upgraded. The key is to figure out how to tune those enzymes to the plastic target, like targeting the spike protein on the COVID-19 virus.

But AlphaFold isn't the only game in town. Months after AlphaFold stunned and delighted the scientific world, Meta AI announced they had conducted a vast survey of genetic material, what it called "metagenomics . . . the 'dark matter' of the protein universe."[11] The company drew the DNA from bacteria, fungi, and viruses from the ocean, soil, and even humans.* The AI angle was that Meta's AI, given a string of amino acids, could predict with precision how they would fold into their final 3-D shape. Meta AI could do it sixty times faster than AlphaFold.

It's a competitive world but also one where really amazing scientific progress is being made. This is a good example.

The Imagined Realized

The genome, the microbiome, AI folding proteins—these are three out of hundreds of active, even intense areas of research, where biology and information technology collaborate. Imagine typing a gene sequence on a computer and sending it to a specially adapted 3-D printer that generates an actual working protein. That's kind of where we're at now. The most stunning thing is the takeover by modern technology of what is surely the fundamental process of life—DNA making proteins. Twenty-five years ago, this technology would have been thought to be not just unlikely but fantastical.

---

* The DNA from this huge range of organisms is sequenced. Once that is done you can tell what parts of this mass of DNA are actually codes for proteins—millions are, and their proteins are unfamiliar to us. Meta AI can predict the final folding shape just from the order of the protein subunits, the amino acids.

CHAPTER 5

The Good Long Life

We are in the midst of revolutionary changes in human longevity. The Population Division of the United Nations estimated there were 23,000 one-hundred-year-olds in the world in 1950, and by 2021 that number was expected to be around 573,000, nearly a twenty-five-fold increase. Centenarians are no longer outliers, as they were in previous centuries. It wasn't that they didn't exist, but they were a minuscule fraction of the population. The vast majority didn't come within shouting distance of age one hundred. Better medical care, nutrition, and a cleaner environment are partly responsible for the gain, which raises other questions, especially "Is there a natural, fixed limit to a human life?" There is a growing belief that there might not be and that has given some scientists visions of not just inching the human life span upward, but launching it into a whole new era.

Would you like to live to 150, or even more? To have great-great-great-great-grandchildren? One of your first reactions might be "Are you kidding? Who would want to live physically and mentally spent for the final fifty years?"

You'd be right . . . unless there was a guarantee you wouldn't be plagued in your extreme old age by the chronic and debilitating diseases that are so common today. This isn't a dilemma you should spend that much time thinking about—extending life is a huge challenge, and adding fifty years? It seems unreal to even contemplate, yet labs around the world are trying to identify what limits the human life span and how to counteract, or even eliminate, those limits.

You can use a conservative approach ("This might be within reach") by trying to change the shape of the curve of human aging from its current long downward slide to the end, worsened by multiple illnesses, to a long-lasting healthier horizontal plateau that suddenly drops off into death. A more radical version ("This might be forever beyond our reach") seeks to extend human life anywhere from twenty, to thirty, fifty, maybe a hundred years, maybe much more.

The two approaches are not independent: some of the data gathered to reduce the impact of disease in late life may turn out to be a tool to extend life.

The Starting Line

A quick guide to human aging: Life expectancy, the age you can expect to live to depending on when you were born, has risen dramatically since the early 1800s. In high-income countries, that increase has been an additional year for every four that pass. My father, born in 1909, had a life expectancy of fifty-five or fifty-six, but my son, born in 1992, could expect to make it to seventy-eight. One year for every four that pass actually boils down—statistically—to a gain of about six hours every day. So a baby born yesterday has a shorter life expectancy by that amount than one born at the same time today. And so on.

Remember, these numbers reflect those who used to die in their early years from childhood infections, especially in the pre-antibiotic era. Even in the mid-twentieth century there were still few effective treatments for cancer or heart disease. Also, the steady increase in life span was not uniformly distributed. Those living in places where the chlorination of water or pasteurization of milk were first established were beneficiaries, as were those where vaccines were available (as we continue to see). And to demonstrate that there are always outliers—my dad came within a month of celebrating his ninety-eighth birthday.

These increases in life expectancy, while robust in appearance, can be slowed at a moment's notice. HIV, the opioid crisis, and of course COVID-19 have all temporarily stalled or even reversed the steady climb in life expectancy. The last time life expectancy worldwide was dramatically reduced was during the 1918–20 influenza pandemic.

But the life expectancy curve still climbs, the number of centenarians is rising at an incredible rate, and so, taken all together, these numbers pose the question: What is the limit of a human life—or is there one?

A Century Is Nothing

The oldest human ever recorded (biblical exaggerations notwithstanding) was a French woman named Jeanne Calment, who died in 1997 at the age of 122. Since her death, Calment has evolved from a natural wonder to a possible fake. When she was born in 1875, the life expectancy for a woman was 45 years. Calment smashed through that barrier: she was still riding her bike at 100, lived alone until she was 110, and quit cigarettes when she was 117. (I thought perhaps she was afraid they would shorten her life, but in fact she couldn't see well enough to light them.) But were those feats real?

In 2018, Russian mathematician Nikolay Zak, encouraged by his geriatrician friend Valery Novoselov, published a speculative, non-peer-reviewed paper in which he argued that Calment wasn't Jeanne Calment at all, but instead her daughter Yvonne.[1] Yvonne was supposed to have died of tuberculosis in 1934. Zak argued it was actually Jeanne who had died in 1934 and Yvonne took on her identity, "in order to avoid paying inheritance tax." Zak left no stone unturned, examining descriptions of Calment to see if her physical appearance had changed after 1934, studying pictures of her, calculating the odds of someone living that long, going to extreme lengths to make his case. However, the furor over Calment's legitimacy has died down since 2018, because Zak's case just wasn't persuasive.

An American named Sarah Knauss died at 119 to take second place, and the oldest person at the time of writing is Kane Tanaka, who just celebrated her 119th birthday and is determined to hit 120.* The next seven oldest, all women, reached 117. Only one is still alive. The oldest man ever was 116-year-old Jiroemon Kimura.

The Gompertz-Makeham law of mortality decrees that after about the age of 30, a human's likelihood of dying doubles every eight years. Yet studies of centenarian (100-year-old) and even supercentenarian (110-year-old) Italians suggest that somewhere around the age of 100 or 105 their risk of death levels out, until each year it's a 50-50 proposition, a toss of a coin. Either you die or you don't. In this view, living to 125 or even more is possible, just like throwing ten heads in a row is possible. (One author has suggested that soon someone will live to 128.) It's rare but it could happen. That living to record ages is a possibility does suggest that there might be no absolute limit to human life, because as the human population grows, and more people live to those ages, another Jeanne Calment could come along if the mortality coin goes on a run.

The question that arises: Do these statistics support the idea that the maximum human life span is around 115 years? And if so, what might be done to push past that limit?

Who Lives the Longest?

The first important observation is the disparity between men and women, and it's not just borne out by these oldest of the old. An independent set of statistics portrays the difference starkly: half of all women over 65 are widows. Widows outnumber widowers 3:1. At age 65 there are 100 women to every 70 men, but by age 85 those numbers are 100 to 38. Ironically, it doesn't start that way. More boys are born than girls, about 104 to 100 at birth, but then the boys start to die slightly more often than girls, especially at puberty, where the mortality rates of boys triple those of girls, mostly because of violent deaths attributable to reckless behavior.

---

* Unfortunately, Tanaka didn't make it to 120. She died at the age of 119 on April 19, 2022. Then the next oldest, Lucile Randon, died in January 2023. At the moment, the oldest person in the world is Maria Branyas, who's nearly 116.

Once the boys lag behind they never recover. Across the globe women live about 5 percent longer than men.

Why do women live longer? Sex hormones are one reason. Cisgender women (whose gender identity matches the biological sex assigned to them) have more estrogen and less testosterone. Both are advantageous. Estrogen affords some protection against heart disease (and maybe cognitive decline, although that's controversial), while testosterone raises the risk of some cancers and contributes to the risky behavior that plagues young men. One classic study showed that from the 1500s to the 1700s, eunuchs in the Imperial Court of the Chosun Dynasty in Korea benefited from their loss of testosterone by living on average twenty years more than comparable uncastrated men in the court.[2] There were records for eighty-one eunuchs in all, three of whom lived to over a hundred. The kings they served didn't live nearly as long.

Behavior plays into the male/female difference, too: men used to smoke and drink much more, but women have been catching up and narrowing the life expectancy gap. But again, the difference in life expectancy is only 5 percent, which isn't really significant if you want to bust through the life expectancy barrier. And really, these sex differences are but a tiny part of the whole aging picture.

We all know people whose biological age seems to be radically different from their chronological age. In the last few years scientists have been getting a grip on some of the factors that might underlie these discrepancies, and it's a complex picture. Known to be important are changes apparent to all of us, collectively known as "frailty," like cognitive status, verbal ability, walking speed, and grip strength, but many are inapparent, only detectable in the lab, like telomere length (the caps on the ends of chromosomes that shorten with every cell division), epigenetics (the process by which over time DNA is festooned with small chemical groups), and a host of changes in other key biochemical reactions. The majority of these changes are better indicators of biological age than the years you've lived. Regardless of the goal, either to shorten the period of ill health prior to death or extend life itself by many years, some of these processes, especially the biochemical ones, might be ready targets.

Obviously, there's still much to be learned from human studies, and it looks like it will take years to learn it. But what can we draw from the aging of other species? Horses live to thirty; cats to their mid-twenties; some dogs to their teens. None of these is extraordinary. But there are organisms that seem to defy aging, and there's a lot of research under way to see how they do it, and more important, to transfer those secrets to human aging. Honestly, the variety of organisms from which we can learn gives new meaning to the phrase "all creatures great and small."

"I Refuse to Die!"

The hydra is a small (one-centimeter or 0.39-inch) pond-dwelling ambush predator, just a tube with tentacles at one end that anchors itself to a plant and waits for prey to swim by. The rattlesnake of the pond. But they fascinate scientists because they appear to be immortal. Cut a hydra in half and it will regenerate two complete hydras.

Their secret is that they're basically bags of stem cells, cells that can be programmed to become a variety of specialized cells, like the lining of the gut, a tentacle, or an ovary. We are definitely not bags of stem cells. However, if one of the ways to delay aging would be to use stem cells to rejuvenate failing tissues and organs (and we're seeing some of that already), then the hydra is the perfect model.

Caenorhabditis elegans, better known as *C. elegans*, is a roundworm that is hugely popular in labs all over the world. In the world of aging research, *C. elegans* proved its worth when Cynthia Kenyon, a scientist specializing in aging, isolated one of its 20,000 genes, called daf2. Worms with the mutant version of daf2 live *twice* as long. Two and a half weeks, which is the equivalent of a two-hundred-year-old human.

So all we have to do is identify the human equivalent of this gene and we're almost there, right? Not really. A roundworm is not close to a human, and most researchers strongly believe that while genes are crucial to aging, no single gene is.

Then there are naked mole rats, odd little mammals that maintain colonies with a single breeding female like honeybees, are extremely resistant to pain, and can go without oxygen for something like eighteen minutes.

But that isn't the limit of their bizarreness. Naked mole rats, like that population of Italian centenarians I mentioned earlier, defy the Gompertz-Makeham law. Their risk of death, which is about 1/10,000 when they're six months old, *never* rises. So they're not exactly like the Italians, whose death risk does rise throughout their lives and only stabilizes when they're extremely old. Based on their mouselike dimensions, you might expect naked mole rats to live as much as six years. But they live past thirty and no one at this moment knows exactly when they'll stop, if ever.

Finally, there's a variety of large animals that make up part of this longevity zoo: the rockfish (150 years), the bowhead whale (200 years), and the Greenland shark (300 years). Little is known of the factors underlying their immense longevity.

These animals illustrate that life spans can dramatically exceed what we might expect. But to extend the human life span is still challenging because the diversity of animals is matched by the diversity of anti-aging mechanisms, and not all of these mechanisms—maybe none—will be applicable to humans.

But the idea that aging itself isn't a fixed number is important. It has given rise to the idea that curing the diseases that are much more common in advanced age, like cardiovascular, cancer, autoimmune conditions, and cognitive decline, won't have nearly the impact on longevity as identifying the independent causes of aging itself. This approach is called "geroscience."

It's been estimated that eliminating cancer might only raise life expectancy by about five years, mostly because the risks of other diseases would continue to rise. What about cancer, heart disease, stroke, and kidney disease? Eliminate *all* of them and you'd add an estimated decade to the average human life. But slow down the aging process itself, and life could be extended anywhere from fifteen to twenty-five years, as the diseases of old age would also be delayed.

Some clues come from those animals with the unusual life spans, but we're gaining more information from more familiar animals like fruit flies and mice. One of their great advantages, besides being very familiar to scientists, is their short life. The effectiveness of experimental efforts to extend those lives can be judged over weeks in the case of fruit flies, or months with mice.

One of the most interesting candidate drugs for extending longevity is rapamycin. In some animals it not only increases life span but also has an impact on a variety of diseases. In mice rapamycin can reduce the incidence of cancer, obesity, cognitive decline, and heart disease. The journal *Science* selected as one of its scientific breakthroughs a 2009 study showing that, in addition to curbing disease, rapamycin extended the life span of mice.[3] Since 2009 there has been a stack of studies reinforcing rapamycin's effect on longevity in mice, one of the most important aspects being that even if given late in life it has the same life-span-increasing effect. There had been concerns that such effects would only be seen if the drug were given early, before significant aging-related damage had occurred.

Also of note, rapamycin has already been approved by the Food and Drug Administration in the US for use in preventing rejection of transplants because it suppresses the immune system. It's not without its dangers, for that very reason. Even so, some doctors in the US are prescribing it to patients who want to delay aging, so-called off-label use. If they think it's warranted, they are free to prescribe a drug for purposes other than those it was approved for. Of course, claims of efficacy coming from such patients aren't usable evidence that it actually works.

Clinical trials would be. One eight-week trial with older patients to test its safety showed that rapamycin was well tolerated, but a full-on trial that could establish life extension, or not, has not yet been conducted. Some aging experts are surprised such trials haven't yet begun. Could that be because extending life seems either impractical or unnecessary?

Tired Blood

It's funny how the many current approaches to aging, like rapamycin, have connections to the old. It's been known since the 1930s that calorie reduction in mice lengthens life—it is now clear that rapamycin has an important role to play in the metabolic connection between the two. Another old idea, a centuries-old belief that never really worked that well, was the idea of joining the circulation of a young animal to an old one, to infuse "young blood" into an old body. There are grotesque tales that in the fifteenth century, Pope Innocent VIII's personal physician bled three young men to feed their blood to the dying pontiff in the hopes of reversing his condition. Like an early and very clumsy version of blood transfusion. Some sources claim that it was bidirectional: "All the blood of the prostrate old man should pass into the veins of a youth who had to yield up his to the Pope."[4] That particular detail is disputed. Regardless, the three youths died, as did the pope.

An equivalent modern procedure called parabiosis has been very successful in a range of modern experiments in mice. So much so that scientists are now hot on the trail of identifying the substance(s) in the blood that could be responsible for literally turning back the clock for old mice receiving transfusions of young blood.

The First 150-Year-Old

How would you feel if you could tack on another twenty or thirty or more years of life? Healthy life at that? Mixed reactions seem to be a fair summary of the answers, but there's a big push to lengthen human life. Where does the impetus for that come from? Believe it or not, one of the hotbeds is Silicon Valley. Cynics have suggested that Silicon Valley is the perfect place to embrace the idea of extending human life by decades: it's nontraditional and there are many individuals superconfident in their own abilities (and therefore desirous of living a very, very long time).*

Many high-quality scientists have been lured there. I've already mentioned Cynthia Kenyon, a well-known aging researcher, professor emerita

---

* If Silicon Valley weren't enough, Saudi Arabia recently announced their own program called the Hevolution Foundation, which will spend $1 billion a year to fund anti-aging research.

at the University of California, San Francisco, and also vice president of aging research at Calico, a Google spin-off. One of the world's foremost experts on those naked mole rats, Rochelle Buffenstein, is also at Calico. Ray Kurzweil, computer scientist and director of engineering at Google, has been very public about his determination to beat back aging.

But scientists aside, these Silicon Valley centers for aging research stand out for two reasons. One, the number of extremely rich people (men) backing them. Amazon founder Jeff Bezos has apparently invested in Altos Labs (billionaire Yuri Milner, who made his money in Facebook, is the founder). PayPal cofounder Peter Thiel has helped fund Unity Biotechnology.* The other notable feature of these Silicon Valley labs is that they're home to some of the most bullish optimists about their ability to conquer aging. Aubrey de Grey is one.

For years de Grey was the chief scientific officer at the SENS [Strategies for Engineered Negligible Senescence] Research Foundation, but late in 2021 he was let go for allegations of sexual misconduct.[5] SENS is both about money and extraordinary claims. The SENS website overwhelms a visitor with invitations for donations, and de Grey argues that we might be a decade away—or even less—from seeing the results of the anti-aging techniques he has been promoting. He liked to refer to something he called "escape velocity," an analogy with rockets escaping the Earth's gravity.

De Grey claims that if people living today are able to gain a 30 percent increase in life expectancy, that is another twenty-five years on top of, say, an age of eightyish, then in those twenty-five years, medicine will provide another 30 percent increase. So what you'd be seeing is that life expectancy, rather than shrinking with age, would be growing. While he's not the only person to make such a statement, he has also said, "The first 1,000-year-old is probably only 5 to 10 years younger than the first 150-year-old." Other enthusiasts have claimed that the first 150-year-old is already alive today![6]

---

* A few years ago, there were rumors floating around that Thiel was interested in investing in the idea of parabiosis, transfusing the blood of young men into older ones.

Ray Kurzweil, predictor extraordinaire, is famous in the anti-aging community for his daring "we'll fix aging by merging with technology" stance. Even though he thinks that possibility lies somewhere in the future, he's not waiting for it to happen. Depending on the account, he's taking anywhere from ninety to two hundred pills a day to stave off decrepitude and claims that his biological age is decades less than his chronological age. But that, according to Kurzweil, is only the beginning. He has come up with the "law of accelerating returns," the claim that not only is technology progressing rapidly but the rate of that progress is rising, too. What might take one hundred years to develop at today's rate or progress will, as a result, actually happen much sooner. Kurzweil argues it applies to a variety of technological developments, including the assault on aging. In fact, he's even more optimistic than de Grey about de Grey's own "escape velocity," arguing that it should happen within fifteen years.[7] This phase will rely on "nanobots," tiny devices that travel through our arteries and veins, seeking out and eliminating developing problems, like incipient tumors, atherosclerotic buildup on the walls of arteries, whatever. This in the 2030s.

The final step, in Kurzweil's mind, will be when robotics, nanotechnology, genetics, and AI come together, and immortality—electronic or electrobiological—will be the norm. I'm not sure Ray has ever claimed that the first 150-year-old (or any other age) is already alive, but I'm sure he would enthusiastically agree.

The connection between longevity improvements and money is perfectly symbolized by Longevity House in Toronto, where a onetime $100,000 fee gets you access to "biohacking, plant medicine, epigenetics, breathwork, and functional medicine to balance mind, body, and spirit."[8] And live much longer.

Even if you don't buy that dream, or those of a clutch of Silicon Valley billionaires or the predictions of the de Greys and Kurzweils, even if you are content to think that the "health span" can be extended into the nineties or even the one hundreds with little disability until death arrives suddenly, it's hard to believe that aging is going to look the same over the next few decades.

Part II

What Will We Eat?

CHAPTER 6

Food Fight

In 1800, 90 percent of the population was involved in farming; today, it's 1 or 2 percent.

To feed the world's population over the next twenty-five years, humanity will have to produce as much food as it has over the last several *thousand years*.

Good news: today, it only takes 30 percent as much land to feed a person as it did five thousand years ago.

Bad news: there are 138 *times* as many people now as there were then.

It seems that no matter how successfully we meet challenge after challenge trying to feed an expanding world population, gains are slow. It's estimated there are about as many malnourished people today as there were ten years ago, despite the incredible strides that have been made in food production. In one sense that's a huge success, given that in that time the global population has grown by 800 million—coincidentally about the same number of malnourished people in the world. The pressure to feed more people is growing, and there are enormous environmental, social, and political challenges standing in the way. The question is, what role(s) can technology play?

Focusing on food production technology runs the risk of over-looking other crucial aspects of food. An example is food waste, which, depending on who's measuring and where, can amount to 50 percent of the food produced. There are losses at every stage before it lands on your table—or really, in your mouth. This amount of waste gives weight to the old idea that we don't have a food shortage, we have a distribution and/or a waste problem.

The EAT-Lancet Commission on Food, Planet, Health has no doubt that reducing food waste substantially "is essential for the global food system to stay within its safe operating space."[1] It's one area where government policy and consumer reaction are going to play significant roles. I don't know where innovation will come from, but I think we might gain some ground there. Whatever we achieve, reducing waste must be accompanied by greater production.

In 1968, ecologist Paul Ehrlich published a book called *The Population Bomb*, which sold an unreal two *million* copies.* But in the years after its publication, the book attracted criticism, partly because of what it said and partly because of how it said it, but mostly because of how wrong it was. (The avowedly conservative *Intercollegiate Review* ranked it as one of the worst books of the twentieth century.)[2]

But the book grabbed everyone's attention because of such bold pronouncements as this one made in the prologue:

The battle to feed all of humanity is over. In the 1970s the world will undergo famines—hundreds of millions of people are going

---

* While Paul Ehrlich appears as the sole author of the book, he later admitted that he and his wife, Anne, shared the writing. However, the publisher preferred that it appear under his name alone. Paul Ehrlich has often said he's embarrassed for having agreed to this.

to starve to death in spite of any crash programs embarked upon now . . . many lives could be saved through dramatic programs to "stretch" the carrying capacity of the earth by increasing food production. But these programs will only provide a stay of execution unless they are accompanied by determined and successful attempts at population control.

What follows are more than two hundred pages expanding on this dire prediction. These include three "scenarios," in-depth timelines of possible futures for an overpopulated and starving world. Two eventually lead to nuclear war; the third, death by starvation of half a billion people. In recent years, Paul Ehrlich and his coauthor, wife Anne Ehrlich, point to the phrase in the book that introduced these scenarios: "Remember these are just possibilities, not predictions . . . but they describe the kinds of disasters that will occur as mankind slips into the famine decades."

That reminder hasn't diminished the criticism. Nothing in the book has really come to pass, and while the authors still contend that the challenges they identified have not been solved, their work has ironically given encouragement to the opposing idea: the more people there are in the world, the more human-powered ingenuity for problem solving we have at our disposal. Call proponents of this view technophiles if you want, but they argue that given that the world population is now 8 billion, and worldwide starvation hasn't transpired on the scale envisaged by the book, we can look optimistically to the future.

That is not the only critique of the Ehrlichs. Others are focused on the implicit racism in their idea that population growth must be halted. Those parts of the world where population is increasing the most also happen to be poorer countries and nonwhite. So it's worthwhile keeping *The Population Bomb* in mind when discussing how to feed the world in the future: that is, be wary of predictions, projections, or scenarios of any kind.

Upending Ehrlich

The Ehrlichs' view of the future was sidelined by an agricultural transformation. Conceived in the 1950s and executed through the 1960s and '70s, the Green Revolution was a multipronged international effort to raise food production dramatically with the rapid development of new and more productive strains of crops; the application of highly efficient mechanized farming; and the intensified use of irrigation, pesticides, and herbicides. The results were spectacular: estimates vary, but yields of the world's food crops, especially wheat, rice, and corn, have risen by more than 300 percent over the last six decades, while at the same time land use for agriculture rose less than a tenth of that. An astounding achievement.

In 1968, Paul Ehrlich had written, "India couldn't possibly feed two hundred million more people by 1980," yet the country became self-sufficient only six years after his dire prediction. A common estimate is that a billion people were saved from death by starvation by the Green Revolution. The American agronomist Norman Borlaug, one of its leaders, was awarded the Nobel Peace Prize in 1970.

But the Green Revolution wasn't perfect. Food wasn't distributed evenly around the world, and production relied on dramatically increasing the use of herbicides, pesticides, irrigation, and fossil fuels (for machinery), all of which have had a negative impact on the environment. It's generally agreed now that what the world needs next is an "Evergreen Revolution." An oft-quoted claim is that by 2050 we will have to *double* the amount of food produced today, even though the global population will likely only increase by about 25 percent, from today's 8 billion to 9.8 billion. I say "only" to contrast the increase in population (25 percent) with the estimated need to increase food production by 100 percent.

Why the disparity between the population increase and the declared need for proportionally greater food production? Why shouldn't we only need 25 percent more food? The 100 percent claim goes back more than a decade to statements made by a committee of the UN General Assembly.[3] Acknowledging the positive impact of the Green Revolution, the committee nonetheless argued that going forward there was a wealth of uncertain factors besides a growing population: changing food preferences, climate change, and ensuring social security and equitable food distribution. Together those factors suggest that food production would have to be increased, not just in step with population growth, but dramatically more.

Subsequent analyses have revised that 100 percent number down to something closer to 50 percent more food by 2050, but even this lower estimate would have to be accompanied by improving access to food, cutting down food waste, and, most important, dramatically reducing the negative impacts of food production on the environment. Agricultural production, together with the unsustainable conversion of forested land to farms, contributes almost a quarter of the world's total greenhouse gas emissions.

It's of course true that no one country can serve as an example for all, but the Netherlands comes close: it has an area double the size of Manhattan devoted to greenhouses, which occupy one-tenth the space of traditional farming. The Dutch are big-league growers of tomatoes, and they can produce twenty times as many potatoes per square meter as a plot of land in Spain. A perfect example of the social impacts on technology: the Dutch experienced a severe famine post World War II, and their agricultural sophistication today was their response.

How Can the World Be Fed?

Genetic technology, whether traditional or leading edge, is sure to be crucial in mounting an Evergreen Revolution. Improved plant breeding was hugely important during the Green Revolution—a wide range of new high-yielding crops were generated by traditional breeding techniques. One example was a wheat called semi-dwarf, the thicker and shorter stalk of which was resistant to high winds, produced more wheat, and was rugged enough to be harvested mechanically. Originally planted in Mexico, it eventually spread around the world.

However, more advanced genetic technologies are now available, as I described in Chapter 4: single genes can be added or subtracted from plants, with new varieties taking a mere two or three years to be ready to plant, versus ten years for traditional breeding. The precision of the new genetic technologies allows for that increased speed; the higher production means that less land is needed to harvest the same number of crops. There's evidence that in the absence of genetic technologies, corn, cotton, soybeans, and canola would have required 60 million more acres of cropland. Breeding has made possible the more efficient use of water, heightened production, hardiness to warming temperatures, and increased resistance to insects and other pathogens.

The potato is a prime example: the pathogen in the dreadful Irish potato famine of the 1840s was an organism called late potato blight. Some versions of it are still around—it is still the costliest potato pathogen to deal with, although the exact one that caused the famine is probably extinct. It likely arose in the Toluca Valley in Mexico in the early 1800s, then made it to Europe by about 1840.[4]

Some potato varieties have genes that give them resistance to this fungus. If twenty years ago you set out to combine that resistance trait with some others, in the same plant, you'd be looking at years of breeding experiments. But recently a team of scientists transplanted three genes for blight resistance from one variety to another one popular in East Africa.[5] Five years of field trials in Uganda and these "genetically engineered" potato varieties are described as being "virtually 100% resistant to late blight." Sounds good so far, though the word *virtually* worries me a little.

The existence of the potato genome practically ensures that many similar improvements to the crop could be made. The genetic switches that control development in the potato are now known. Using them might make it possible to develop new strains with a greater tolerance for long growing days. Potatoes are native to the Andes and depend on short days to stimulate growth. Heightening the tolerance for daylight could allow farmers to extend the potato's range.

In the future, climate change will be the inevitable but uncertain factor dictating how and where food production can be maintained or, ideally, increased. Climate change doesn't just affect how well familiar crops will fare in a changed landscape, but what happens to their pollinators, too. At the moment, they're not doing well. What is certain is that a future map of prime agricultural production around the world is likely to look very different in 2050 than it does now. Genetic technology will be used to improve heat resistance in warming areas, to adapt for transport to areas that were formerly too cool to be productive, or even to modify crops for so-called protected environments, either greenhouses or vertical farms.

Grow Up

Vertical farming is the most radical of these "protected environments" and is a perfect example of the difficulty of prediction. In one stroke it aims to reduce dramatically the transformation of natural landscapes by taking farming indoors and growing crops in stacked trays. Housing such setups isn't difficult: an empty warehouse or a shipping container could replace a field. It may sound like this is much too small-scale to make any sort of dent in future global food needs; it might also run into difficult technological issues. But there is much about vertical farming that, at least at first glance, makes sense. Growing indoors obviously enables much greater control: no drought, no flooding, no crop-damaging hail, and the possibility of better control over crop-killing pests. Artificial lighting and temperature control allow year-round crops, and a vertical farm can be put in places where traditional farming isn't possible. Vertical farming also provides an opportunity for significant water savings, because water can be recycled and evaporation reduced.

Of course, greenhouses are a precedent for vertical farms, but even here the vertical arrangement has an advantage: greenhouses tend to be located far away from urban centers (where much of the food is consumed) because land is much cheaper there. The time and cost involved in shipping produce to the city often forces farmers to harvest a crop before it's fully ripe. By stacking greenhouses, production could be brought closer to the city and reduce the need for a too-early harvest.

Taking the concept even further, plants could be grown hydroponically (without soil), where they are bathed in nutrient solutions with roots supported by synthetic materials, or even aeroponically, where the roots are sprayed with a water-based solution containing essential nutrients.

The list of apparent advantages seems endless: LED lights could be controlled to emit the intensity and duration of growth-enhancing illumination and tuned to the ideal part of the visual spectrum. A typical claim is that a city-lot-sized vertical farm can produce as much as a sixteen-acre farm while using a tiny fraction of the normal amounts of water.

Yet there are vocal critics who argue that vertical farming is just another example of a technology whose deployment would fall far short of easing food insecurity around the world and would likely cost too much to build and too much to run. The criticisms of energy consumption are hard to evaluate because the costs of renewable energy have come down so quickly that a critique written years ago may no longer be relevant. For instance, in a 2010 column, the *Guardian*'s George Monbiot was scathing in his dismissal of vertical farming, arguing that the cost of solar energy to power a vertical farm would dramatically exceed growing the same crop on a traditional farm, calling the idea, in his usual soft-spoken way, a "green delusion" and "towering lunacy."[6]

It is true that collecting and distributing solar energy that could provide enough "sunlight" is challenging, but Monbiot's column was a long time ago in the life of solar energy. Regardless of the exact costs, food scientist and author Stan Cox has pointed out the irony that vertical farms would use solar energy to generate electricity that would then be used to power lamps that would convert that solar energy into artificial sunlight.

The most recently relevant scientific analysis studied the costs associated with vertically farmed wheat. The conclusion was, "Yields for wheat grown in indoor vertical farms under optimized growing conditions would be several hundred times higher than yields in the field due to higher yields, several harvests per year, and vertically stacked layers. Wheat grown indoors would use less land than field-grown wheat, be independent of climate, reuse most water, exclude pests and diseases, and have no nutrient losses to the environment." All good, right? Then the punch line: "However, given the high energy costs for artificial lighting and capital costs, it is unlikely to be economically competitive with current market prices."[7]

That is why there are very few vertical farms that are economically successful and most of them do not produce the kind of crops that are likely to be instrumental in combating global hunger. Those would include wheat, soy, and corn, but right now vertical farming is confined largely to salad greens. This isn't to say that vertical farms won't play a role, but my feeling is that it will not be as significant as less glamorous solutions like more equitable distribution of food and the genetic improvement of yields.

It Won't Be *This* Bad

I began this chapter with Paul and Anne Ehrlich's *The Population Bomb*, the book that predicted horrific consequences for a world in desperate need of food. Theirs wasn't the only look at an undernourished, overpopulated future. Another, the 1973 movie *Soylent Green*, was set in the year 2022. New York City's population was 40 million and only the elite could eat natural food. Everyone else was on a diet of processed wafers of different colors: red, yellow, and green. The green were the tastiest.

Over the course of investigating the murder of a prominent New Yorker, Detective Robert Thorn finds out that the green wafers might have an unusual, secret, and disturbing origin. Sure enough, he tracks their origin to a waste disposal plant and reveals to a crowd in the last line of the movie, "Soylent green is people!"

We're not there yet.

CHAPTER 7

The Issue Is Meat

There's no technology in the world that operates independently of human preferences. Meat is a good example. Over the last few decades, meat production has set records that could hardly have been predicted. A lot of environmental harm has accrued along the way. Meat production has to do more than just keep pace with the growth in population; it also has to keep pace with changing tastes, and, as incomes rise, more and more people have started to consume more and more meat.

[T]he more highly prized varieties of animal foods—such, for example, as beefsteak or chicken's breast—will be grown in suitable media in the laboratory. It will no longer be necessary to go to the extravagant length of rearing a bullock in order to eat its steak. . . . So long as the "parent" is supplied with the correct chemical nourishment, it will continue to grow indefinitely and, perhaps, eternally.

—Earl of Birkenhead, 1930[1]

The issue of where meat might come from certainly wasn't front and center when the Earl of Birkenhead (Frederick Edwin Smith) published his book *The World in 2030 AD* nearly a hundred years ago, but it is today.* Most current discussions conclude that we simply can't continue to do what we've been doing: keep increasing the production of meat to keep up with demand from a growing urban world. Meat grabs a bigger share of the diet as income rises.

You may be thinking steaks and ribs, right? But no, articles in the journal *Meat Science* commonly group cattle, pigs, and chickens together as "meat." One article showed that from the early 1960s to 2011, annual meat consumption around the world shot up from 23.1 kilos to 42.2 kilos, or from 50.9 pounds to 93 pounds—not quite doubling but pretty amazing nonetheless.[2] The striking thing was that as the gross domestic product (GDP) of countries and/or personal income rose, the consumption of meat rose with it. Other studies have pegged a US$36,000 annual salary as a rough threshold for meat consumption to rise.[3]** There is also evidence that somewhere around US$55,000, the connection between meat consumption and wages is lost—salaries continue to rise but meat consumption steadies. It's impossible to predict how many countries will reach the GDP level that triggers the increase in meat consumption or the higher level where they disconnect, but some say we'll need 50 to 70 percent more meat, which is daunting, even if production has quadrupled since the 1960s.

Even Bigger Numbers

Of course optimists focus on one set of numbers, like those reflecting production above, but look at a different set and it's harder to remain positive. For instance, of the total earthly biomass, domesticated animals

---

* The Earl of Birkenhead's book is quite incredible. He deserves praise as an unsung futurist, predicting as he does wind and tidal power, stereo television in full color broadcasting live from anywhere in the world, and a deep understanding of genetics. Of course he makes some errors, but he also added a plea for something pleasant beyond caffeine, alcohol, and tobacco.

** You'd imagine that number is higher with global inflation.

make up 60 percent of mammals on Earth; the rest are wild. The vast majority are cattle, pigs, and sheep. Next in line: humans! We represent 36 percent. So domesticated animals and humans together make up 96 percent of the world's biomass: all wildlife is just *4 percent* of the total. Think of everything you've seen on all those David Attenborough documentaries—all that incredible wildlife diversity—completely swamped by barnyard animals.

Today there are more than a billion cattle worldwide, roughly the same number of sheep, and somewhere between 700 to 800 million pigs.* While these numbers do fluctuate from year to year, they have risen dramatically over the last century. The numbers we see today underline the fact that vast numbers of wild animals have been displaced by the domesticated versions. And underlining the rise in domesticated animals are the demands they put on the environment.

Cattle emit nearly 100 kilograms or 220 pounds of greenhouse gases for each kilogram of meat. Poultry emit just one-tenth as much (10 kg/ 22 lbs), and pigs a little more than that (12 kg/26.4 lbs).[4]

Huge amounts of land are required either to grow feed or to allow the cattle to graze. They provide 18 percent of the global food supply but occupy 77 percent of all agricultural land. What was called "clearing the land" a hundred years ago we'd now call deforestation.

With more cattle come more environmental problems directly attributable to them, like extra greenhouse gases and nitrogen released in manure; even more greenhouse gases released by the cultivation of cattle feed; and the conversion of natural landscapes to farmland inevitably reduces biodiversity *and* adds yet more greenhouse gases to the atmosphere.

Climate scientists use threshold temperatures to gauge the progress of climate change and create goals. No more than 1.5 degrees Celsius above pre-industrial temperatures is one. Two degrees Celsius is a more concerning rise that many fear we will reach. Some scientists are so

---

* An old-school museum diorama entitled "Mammals of the World 2020s" might show eight cows, a small herd of sheep, a scattering of pigs, and in the foreground, a field mouse to represent wildlife.

concerned about the hefty contribution of livestock to climate change that they have gone so far as to advocate for the complete *elimination* of livestock—that would reduce emissions by half the amount needed to stay under the 2-degree goal.[5] In addition to the cessation of bovine greenhouse emissions, eliminating meat production would also allow farmland to revert to lower-emission meadows and forests, which in turn could act as potential carbon sinks.

Seriously—What Can Be Done?

At first glance the numbers are damning: livestock occupy so much land and consume 6 billion metric tons of feed every year. That's one-third of global cereal production. Yes, the numbers are ridiculously high, but there are also counterestimates that a large percentage of that feed (maybe as much as 80 percent or more) is not "currently" used as human food, and a lot of the land occupied for grazing is unsuitable for growing human food.

But cattle are *very* unlikely to be eliminated entirely, so we need another, more practical strategy. One is genetic: use CRISPR to insert genes into meat-producing animals—and fish—that stimulate muscle growth without increasing the amount of feed. This approach is in the experimental stage and needs a lot of fine-tuning before you'll see it on the meat counter. Another idea is to feed livestock *entirely* on the plant material that humans don't already eat. This would include grazing on grasslands, or processing waste materials and by-products of food production into cattle feed. There would be multiple benefits to confining cattle to grasslands. Plowing those grasslands almost immediately releases half the carbon they store, so having cattle graze them instead sounds better, but additional claims that cattle grazing on grasslands promotes the storage of carbon by stimulating new plant growth are not borne out by the research so far.

In the end, no matter what approach is taken to raising livestock, it's becoming clear that reducing the consumption of meat must play a role in guaranteeing adequate global food supplies, while at the same time maintaining a healthy ecosystem. CRISPR-enhanced meat might play some role, but there are other opportunities for dramatic technological changes.

The Agriverse

Three technological approaches to reducing the environmental stresses associated with raising hundreds of millions of animals to feed us are the production of plant-based meat, lab-grown meat, and dramatically ramping up the consumption of insects.

Plant-based meats, the best-known being Beyond Meat and the Impossible Burger, have attracted the most attention, but the still-unanswered question is how much positive environmental impact they will have globally. The potential is there: a study done for Impossible Foods made the claim that in comparison to ground beef the Impossible Burger uses 96 percent less land, generates 89 percent less greenhouse gases, and uses 87 percent less water.[6] Those are fantastic numbers, but to have a global impact, plant-based meats have to be more than just expensive alternatives in developed countries.

Impossible Foods differentiated itself from Beyond Meat and others by adding heme protein to their products, a version of heme from soy (although in this case produced in yeast) that gives meat its red color and, according to Impossible Foods, the taste and smell of beef, too. In a series of experiments, rats were fed 100 times the amount of this heme than the greatest devotee of their burgers would ever eat, each day for twenty-eight days, with no ill effects. Ironically, the addition of heme to encourage meat-eaters stumbled because their heme-generating yeast is actually genetically modified, making this environment-saving food a genetically modified organism (GMO), a fact guaranteed to generate cognitive dissonance among some potential consumers.

After an explosive start, plant-based meat companies in North America stalled seriously in 2021, a trend that may or may not be reliable, given that it was a pandemic year with rising inflation. But the excitement has definitely died down while their list of products is expanding.

Cost might be one of the issues slowing the growth of these companies. Estimates range up to 40 percent more expensive—and even higher—than their meat counterparts. One study found "chicken" nuggets to be double the price of their authentic cousins.[7] This is in spite of the fact that raw plant materials are dramatically cheaper than meat, especially beef. But the plant alternatives require extra processing and the addition of ingredients that account for more than 90 percent of their retail price.

They can be criticized on health grounds as well: plant-based burgers have high levels of sodium (often a third of your recommended daily intake), are highly processed, and some have added sugars. On the other hand, they're good sources of fiber and have much less saturated fat than meat burgers.

I can't help but feel that one of the mistakes was to try to make plant-based products as much like real meat as possible, when it's likely never going to be possible to do so. Fewer of the true meat lovers than hoped seem to be switching. However, it is true that plant-based substitutes have made inroads in the market and in doing so have partially overcome resistance based on the feeling that they just don't look or feel like meat. Where the market ceiling for such food is hard to estimate right now: I wonder if Patrick Brown, the president of Impossible Foods, still holds to the goal he set in a 2021 interview with the *Guardian* when he said, "I want to put the animal agriculture industry out of business. It's that simple. The goal is not because I have any ill will toward the people who work in that industry, but because it is the most destructive industry on Earth."[8] It seems clear that while plant-based protein will make inroads into meat-eating, it will not eliminate it, not on its own.

You Want Fries with Those Stem Cells?

Although plant-based meats were out of the starting gate first, lab-based meats are now being recognized as a viable alternative. Lab-grown meat capitalizes on century-old lab technology—tissue culture, or the ability to grow living cells on a petri plate or in a bottle fed by a nutrient solution.

But while the fundamentals are the same, growing meat "in vitro" is a different challenge. For one thing, little work has been done on growing chicken or cattle cells in culture; the vast majority of culturing has been human, rat, and mouse cells for medical application or research. The second issue is volume; one estimate is that adequate industrial production of animal muscle cells (meat) would need something like a five-story structure to house vessels in which muscle cells are dividing.

In their vats of nutrient broth, these cells need some sort of scaffold to grow on, so that muscle cell fibers will elongate and join together. You know you're dealing with trials of different scaffolds for lab-grown meat when you read statements like "Different constructs displayed a Young's modulus in the same range as the native bovine muscle." (Young's modulus is a measure of the stretchability of a material.) Ideally, the scaffold would be edible, eliminating the tricky process of separating the muscle cells from it; it must also encourage those muscle cells to stick to it and provide adequate exposure to oxygen and nutrients. Already early prototype lab cultures using soy-based scaffolds have been described as having a "pleasant meaty flavor." The issues of flavor, odor, and texture are paramount given our species' long and close relationship with eating meat. In 2013, Dutch stem cell researcher Mark Post presented the world's first hamburger made with lab-grown meat at a media event in London. It contained ten thousand strips of myotubes—the precursors of mature muscle fibers—beet juice for color, bread crumbs, a binder for texture, and saffron and caramel to amp up the flavor.[9]

All of that drove the cost of this singular burger to more than $300,000, but it made the point that a hamburger could be made without killing an animal. Two taste testers thought the burger tasted good, but they missed the mouthfeel of fat (which hadn't yet been incorporated).

The missing fat is apparently no longer an issue. Spanish company Novameat has gone beyond ground beef to produce steaks from lab-grown meat by extruding muscle fibers only about 100 microns (4/1,000 inches) in diameter wrapped around and through fibers of fat.

Commercially, lab-grown meat is on the rise: the American company Eat Just's lab-grown chicken nuggets have been approved for sale in

Singapore, the first lab-grown meat to be available to consumers. The company admits they will be expensive but promises costs will come down as the process is refined.

Meat That's Green

I should qualify this progress by pointing out that a real steak is a complex arrangement of membranes, filaments, cell types, and even blood vessels, and no one has come close to being able to reproduce it. But there remains the powerful environmental justification for growing meat in a lab (although it brings other benefits, such as the vanishingly small likelihood of pathogens, given that it's produced under sterile conditions). Just as Impossible Foods claims enormous reductions in water and land use and emissions of greenhouse gases for their plant-based products, lab-grown meat companies argue the same.

There remain significant challenges, one of which is the growth medium bathing the cells. Today that medium is used mostly for medical applications, which are much smaller scale than what's needed for the production of food. Also today's cost, about $1,000 a liter or 33.8 ounces, has to be reduced to something in the range of $3 to $4 a liter. In addition, fetal calf serum is a key ingredient but cannot be part of the formula if the goal is to reduce the number of cattle.

Doubts have been raised about claims of reducing greenhouse gas emissions. The issue is the type of gas. Lab-grown meat will avoid the release of methane, which cows burp in considerable quantities. That's a plus because methane is a more potent greenhouse gas than carbon dioxide, but much shorter-lived, exerting its greenhouse effects for only about twelve years. CO_2, by contrast, endures for up to a thousand years.

One long-range analysis found that under some circumstances, lab-grown meat production facilities would end up raising greenhouse gas levels more than traditional farming, because of their carbon dioxide emissions. In this scenario the advantage of reduced methane fizzles out because atmospheric methane levels stabilize relatively quickly. At that point, the carbon dioxide emissions from growing meat in the lab overtake the farm.[10]

However, and there's a significant however, this modeling assumed that lab-grown meat facilities would continue to be powered by fossil fuels, which is a highly unlikely possibility. And because there just aren't any full-scale lab meat factories yet, the researchers had to guess what sort of emissions might be generated. All in all, I don't see this cautionary note as a game changer. The bigger question is, are we going to like lab-grown meat?

Who's a Good Consumer?

There are hundreds of millions of households worldwide where the meat-eating is shared by the parents, the children, the dog, and the cat. In North America, of course, it's over-the-top: a paper titled "Environmental Impacts of Food Consumption by Dogs and Cats" published in 2017 in the journal *PLOS One* pointed out that in the United States dogs and cats consume 20 percent of the food energy that humans do and *a full third* of the energy supplied by meat and meat products.[11] Yet they're outnumbered by humans 2:1.* After laying out the dismal environmental impact of pet food, the author of this analysis, Gregory Okin, professor of geography at UCLA, maybe overreached a little: "Reducing the rate of dog and cat ownership, perhaps in favor of other, less energy-intensive, pets that offer similar health, social, and emotional benefits, would considerably reduce America's overall livestock-related environmental impacts."

Yes, it would. How long might it take to persuade the close to one hundred million American owners of dogs and cats to give up their pets? Too long. Is this an opening for lab-grown pet food?

A company called Because, Animals is on its way to providing lab-grown meat for dogs and cats, what they're calling "cultured meat." It makes sense to me: watching the way my dog eats breakfast convinced me long ago that he'd eat almost anything with enthusiasm. Whether

---

* They also produce 30 percent of the amount of feces excreted by humans.

the meat is from a carcass or was grown in the lab is irrelevant to him. Because, Animals' dog food is rabbit meat obtained by a small draw of blood from a rabbit. The rabbit survives. So do the pluripotent (totally versatile) stem cells extracted from the blood. Masses of rabbit stem cells certainly won't have much of a structure, but that doesn't matter if the goal is to make a pâté or stew for a dog.*

Whatever your resistance might be to this idea, contrast it with the way cat food is produced now. A fair estimate is that cat food is roughly 50 percent from animal parts humans won't eat and 50 percent "fallen" animals—animals not admissible to the human food chain. There are other random objects, like hides and feathers. The whole mess is sent for rendering, where what would otherwise be wasted is turned into "useful" products, like cat food.

Shannon Falconer, CEO of Because, Animals, points out, "In the absence of the pet food industry, animal agriculture could not exist. The industry, in addition to being unable to sell all the rendered meat, would have to pay for it to be disposed as biohazardous waste. All the landfills in the U.S. would be full of rotting material in just four years."

That's one of the barriers to commercializing cultured pet food—there's already a tightly managed, well-financed system. But people's hesitance about lab-grown meat, even when they're not eating it themselves, will be the consumer challenge. Yet the environmental benefit if this took off? As long as the energy requirements of a cultured-meat-producing facility are reasonable, it's a benefit and, if accepted, an animal welfare victory as well.

There is a weirder side to the consumer acceptance of lab-grown meat. In an article in *Wired* magazine, American futurist Amy Webb wondered if, because lab-grown meat requires at most a biopsy-sized tissue sample from an animal, some gourmets might want to try something a little more exotic than beef or chicken, like dolphin or chimpanzee (although it was the phrase "cocker spaniel kebabs" that caught my eye).[12] Webb suggests regulation would just encourage a black market. The legitimate side of

---

* Mice will provide meat for cat food, obtained the same way.

that argument is that because the growth of lab-based meat is under our control, the opportunity for adjusting taste, texture, nutrients—virtually anything—is there. In fact, given that complete genomes of extinct animals exist, it might be possible to engineer a mammoth-meat steak or passenger-pigeon nuggets.

An Oft-Ignored Downside

If the combination of plant-based and lab-based meats were to explode in popularity, there would be huge impacts on farmers around the world and even on farm animals. What if farm animals were to disappear (as the scientists I referred to earlier advocate)? We've been farming animals for millennia and you could argue that it's part of our nature. Some even claim that farm animals benefit from our raising them because if we didn't they wouldn't exist. I like the counterargument that this would only be valid if their lives were worth living, and it's unlikely that animals in factory farm settings satisfy that criterion.

It's too early to predict which plant-based or lab-grown meats will become the most popular, and whether either will become significant. I'd guess that each will become a piece of the puzzle. But Patrick Brown of Impossible Foods, again, has no doubt which will prevail, and it won't be lab-grown meat: "It is never going to be a thing. It misses the real opportunity when you're thinking about replacing animals in the food system," he says. "As we learn what consumers prefer in terms of flavors and textures, we can dial those up and down. You can't do that when you're stuck with whatever an animal cell can do."[13]

I think Brown might be underestimating the ingenuity of biochemists—even if he is one (or maybe *because* he is one).

It's Not a Feature—It's a Bug

I can't leave the subject of meat substitutes without talking about insects. We're moving along some kind of scale of acceptance here: I'd guess most of you would eat or have already eaten a plant-based form of meat, like a burger. It's safe to say none of you have yet to try a lab-grown burger, but you can still answer the question: Would you? And then, what about insects? I've eaten both whole insects—once—and ground-up insects several times. Ground-up is better, but then I'm a North American and I'm not used to them. There isn't much insect-eating going on around me.

But that's not true of the rest of the world. The numbers might surprise you: close to two thousand species of insects are eaten routinely by two billion people. This includes hundreds of species of, in order of preference, beetles, butterflies and moths (in their larval, caterpillar form), ants, grasshoppers, and crickets. Some of the A-list insects are the house cricket, the desert locust, the black soldier fly, and even cicadas.

Insects are touted as being an important source of protein as we move toward 2050 looking for more food and less environmental impact. As a food source they have a lot going for them: you could harvest ten times as much insect protein as you could beef from a given amount of grass. Put another way, it takes 2 kilos or 4.4 pounds of feed to produce 1 kilo or 2.2 pounds of weight gain in an insect, a ratio far better than any animal meat. And 80 percent of the insect is edible versus 40 percent of the cow.

Insects offer equivalent amounts of protein to beef or pork, and it's complete protein (all nine essential amino acids). They have more iron per gram than beef, less saturated fat, substantial amounts of fiber (from the chitinous exoskeleton), and a wide range of vitamins. Their greenhouse gas emissions are about a factor of 100 less than those of cattle. From a nutritionist's or environmentalist's standpoint, if you wanted to concoct an ideal food, insects would probably come the closest of anything found in the natural world.

While the phrase "factory farming" conjures up images of suffering animals packed together to be slaughtered, industrial-scale production might be the only route to making insects a significant supplement to the global diet. For one thing, harvesting them from the wild could drive populations into scarcity and terminate the experiment prematurely. The other issue is that this is already happening! Insect populations worldwide are declining to the degree that experts have called it an "insect apocalypse," blaming a combination of intensive agriculture, including deforestation; heavy pesticide use; and climate change. The loss of insect pollinators is worrisome, and harvesting insects for food would only worsen an already dangerous situation. It's not just the loss of insects that's troubling: the birds and animals that feed on them are being affected, too. What is an anteater to do if there are no ants? The answer is to "domesticate" a selection of insects and there are already flourishing examples of how to farm insects, both for human consumption directly, or the indirect value of using insects to feed animals.*

In East Africa one of the popular insects for protein production is the black soldier fly.[14] Putting together the production of this insect from a selection of nine farms is calculated to produce 9,000 tonnes or 10,000 tons of protein annually. That amount substitutes for enough soy and fish meal to feed 4.7 million chickens. Follow those numbers through for the entire African continent and the benefits are staggering: 54 million tonnes or 60 million tons of insect-based animal food annually, creating 15 million jobs and eliminating 78 million tonnes or 86 million tons of carbon dioxide emissions. In an indirect way, it also creates food for humans: in Kenya alone, insect protein as feed could generate hundreds of millions of dollars for the poultry industry, freeing up enough fish and corn to feed three or four million people.

If all that weren't enough, the waste from producing vast amounts of insects can benefit plants, too. Insects generate waste, mostly poop and exoskeletons left behind by molting. The poop is rich in nitrogen, and the chitin can be metabolized by soil microbes whose populations increase soil health. This might be a recycling opportunity.

---

* Farming insects en masse could run into objections given that there is now evidence that insect brains have at least part of the system necessary for the perception of pain.

You Wouldn't?

The problem with all this is captured by this excerpt from an article in the *Journal of Future Foods*: "While some edible insect species, like grasshoppers and locusts, require the removal of legs and wings prior to consumption . . ."[15] It's true the legs do get caught in your teeth. I've had that happen and there's a moment, just when you realize that a grasshopper's leg is stuck behind your first molar, that it either goes well or very badly. If you devote even a moment's attention to the situation, you're probably lost—revulsion takes over. "Swallow and move on" is my recommendation.

By the way, the good news is you can eat several species *whole*. And there it is: if you think that cultural attitudes might stall the acceptance of lab-grown meat, how about fried crickets?

The idea of eating insects is disgusting to many, but it's not a "human" characteristic—it's cultural. North Americans and Europeans are especially averse to chowing down on insects. There might be historical reasons for this. Humans have domesticated fourteen large terrestrial mammals, thirteen of which are Eurasian (the only exception is the South American llama). So there was no pressing need to harvest insects. Also, although the tropics are definitely not the only place where insect-eating is practiced, tropical insects tend to be bigger, congregate in larger numbers, and can be found year-round. It might be that we came to associate insects with crop-devastating plagues. Almost certainly we came to associate them with disease and decay (flies, anyone?). And, of course, Western peoples have condemned insect-eating as "primitive" in the past.

The aversion to eating insects is related to, but quite separate from or in addition to, attitudes toward insects. People everywhere generally love butterflies, admire dragonflies, and are amused by ladybugs. Blackflies, deerflies, horseflies, and wasps more often inspire annoyance, irritation, frustration, fear, even anger. Anything that stings creates anxiety. But these are all reasonable responses, whereas the immediate disgust I feel if I see a cockroach scuttling across the floor is something different. I mean, the very word *scuttling*! The only threat the cockroach poses is to my peace of mind. Add to that irrational response the stomach-turning thought of biting into the fat abdomen of a caterpillar and I am a difficult person to convert to entomophagy. Apparently I'm not the only one. Dror Tamir, CEO of Hargol FoodTech in Israel, has said, "Having the grasshoppers in front of the package is not a good thing for us."[16]

If insects are to take their place as a significant *global* food commodity, that attitude must change. There are promising ways to achieve that. For one thing, grind up those multi-legged crunchy bodies and make them into flour that can then be used for cookies, crackers, samosas, bread, you name it. Render the disgusting bits invisible. That is likely the most sensible way to go, at least until attitudes change.*

But can attitudes be changed?

In 2015, entomologist Molly Keck at Texas A&M University hosted a thirty-five-dollar-a-plate Bug Banquet in San Antonio.[17] Here's the menu: fire ant queso dip, candied pear salad greens with roasted mealworms, goat cheese quesadillas with tortillas made with cricket flour, and baked apples with cricket granola. Seventy people attended, although how many minds were changed is unclear. A hint that there's still a way to go was revealed in comments like "You couldn't really taste the insects in some of the dishes." But appealing to foodies is a strategy that some think is worth trying. Most of us forget that back in the 1960s, raw fish—sushi—went from small-scale and underappreciated to the ultracommon food it is today. Some of those who are for insect consumption argue that sushi bars and positive publicity did the trick and the same could be applied to insects. This opinion is debated, though: a counterargument is that sushi fit nicely into already-established Japanese cuisine, whereas today there is *no* cuisine that would provide a comfortable fit for insects.

On the other hand, maybe we just stop disguising them. There was, after all, a locust cookbook published in 2004 called *Cooking with Sky Prawns.*

---

* There's no reason insects couldn't be added to pet food as well. And there are several European companies making burger patties with insects (ground up!).

It's worth remembering that we already consume insects on a regular basis. A stunning report was released in 2022: as part of a large-scale effort to categorize how insect populations change over time, a team of scientists at Trier University in Germany analyzed a single tea bag and found DNA evidence of four hundred different insect species. In *one* tea bag![18]

We routinely eat insects in the food we buy at the grocery store. In Canada as many as five dead mites are allowed for every piece of cheese 2.5 by 2.5 by 0.6 centimeters, or 0.98 by 0.98 by 0.23 inches. (No live mites are allowed, though.) You can have as many as ten maggots—but no more—and twenty-five insect "fragments" in every 100 grams or .035 ounces of mushrooms. So you've probably already taken that first step toward full-on entomophagy.

A final reminder: October 23 is World Edible Insect Day.

CHAPTER 8

The Banana and the Broiler

Some of our favorite foods exemplify the bind we can get into as we pursue greater production, ease of handling, and appeal to consumers, all to satisfy the global demand for food. This has happened with an unlikely pair: the broiler chicken and the banana. The most common banana has been bred for many things, but is still susceptible to infectious disease, while the chicken, having set some sort of record for achieving stupendous growth in record time, now seems to be at a crossroads where even greater production raises serious environmental and animal welfare issues.

The common supermarket banana: different bananas would qualify for that title in different parts of the world, but chances are, if you're living anywhere north of the equator and are shopping for bananas, you're likely getting a variety called the Cavendish, named after William Cavendish, sixth Duke of Devonshire, who received a shipment of them from Mauritius in the 1830s. While there are several hundred varieties of bananas in the world, the Cavendish dominates: it is responsible for almost 50 percent of global sales, approximately

50 billion tonnes or 45 billion tons of bananas a year. Even so, it's not the only banana: India and China harvest the most globally, and the hundreds of varieties produced in those countries are mostly for the local market.

The Cavendish is dominant now, but it hasn't been for long. It was really only in the 1950s and '60s when the then-dominant type, the Gros Michel, struck by a fungus called Panama disease Race 1, was unable to sustain production on a global scale, and the Cavendish replaced it. You can still find the Gros Michel in some places the fungus hasn't yet reached and if you get a chance, try it (I haven't); it's supposed to be creamier, sweeter, an all-around better banana. Of course, it can't match the Cavendish when it comes to the most important features, like shorter stems that resist breaking in transit or being damaged by storms. In fact, both versions share similar banana-producer-oriented advantages that helped them make it to the top: no crunchy seeds, highly productive, and shippable without refrigeration.

But that "no seeds" feature, while it guarantees consumer acceptance, makes the plant susceptible to disease. New Cavendish plants appear as suckers on a mature tree and these suckers, which are easily transplanted, are genetically identical to the parent plant. Considering this is how all Cavendish plants reproduce, you can see that the vast banana plantations are woefully uniform genetically, and if a pathogen happens along, as it did to Ireland's potato crops in the nineteenth century, the entire crop is susceptible.*

We're not at risk like the Irish were: it's estimated that by the 1840s nearly half the Irish population was hooked on potatoes. They were the main—and often only—choice on the menu. When the quasi-fungus *Phytophthora infestans* arrived, it ran roughshod over the Irish potato fields, which were susceptible, like the Cavendish banana, by having

---

* Genetically susceptible, but successful: as biologist Rob Dunn has pointed out, if a stand of 37,000 aspen trees in Utah can be considered the largest living organism today, by virtue of their genetic identity, the banana plantations in Central America in the 1950s could likely have claimed the same title then.

inadequate genetic variability. (Inadequate British government efforts to come to Ireland's aid didn't help.)

Vulnerability to an invader is heightened if the target plant is genetically uniform. The fungus causing Panama disease, on the other hand, isn't so constrained: it is continuously generating genetic variants with differing abilities to avoid the banana's defenses and trigger the disease. If the banana is not able to respond by producing its own variants to negate the new versions of the fungus, then it is much more vulnerable. It doesn't help that the fungus is perniciously persistent in soil and can easily be transported on boots, hands, or even the machetes used to harvest the bananas.

Genetics!

Can technology rescue the Cavendish banana? A team led by agricultural scientist James Dale at Queensland University of Technology in Brisbane, Australia, is using genetic technology to enhance the banana's inadequate genome. In a set of trials, Dale's team found they could insert resistance genes into Cavendish plants, conferring significant resistance to the fungus.[1]

That was a huge step toward preserving the Cavendish, but it was at this point that consumer preferences loomed large. One of the genes inserted into the banana plants came from a roundworm. Even though it wasn't the actual roundworm but merely some of its DNA, it turned off anti-GMO banana eaters. However, Dale's team also happily discovered that Cavendish bananas already have their own versions of this disease-resistance gene, although it's usually not expressed strongly enough to exert an effect. The team is now beginning experiments to use the incredibly versatile CRISPR-Cas9 system to enhance the activity of these "home-grown" fungus-resistance genes already present in the plants.

These genetic manipulations are under way, and it seems likely that they will work and we'll have Cavendish bananas for a while yet. But forever? Not likely. The fact that consumers in some parts of the world do not want seeds in their bananas means the Cavendish will have to continue to rely on clones derived from living plants, and its lack of genetic variability will continue to be an issue. There are other pathogens standing by, like black sigatoka, banana bunchy top virus, and banana bacterial wilt, all of which Cavendish are susceptible to. But at least now not just bananas but other crops can be modified for disease resistance and other qualities like survival in the dry conditions and higher temperatures of global warming.

In an optimistic world we could also hope for genetics to help address some of the other issues with banana production. Costa Rica is one of the world's major banana exporters—one out of every ten bananas eaten somewhere on Earth comes from there. Runoff and waste from plantations have been blamed for significant damage to Costa Rica's coral reefs, and the World Wildlife Fund identifies the waste from banana production—bags, strings to tie them together, containers—as the highest of *any kind* of agricultural production.[2]

Of course, genetics is not going to be the entire answer: corporate responsibility, government action, and citizen awareness will all likely be more important, but genetics does offer possibilities that weren't there during the Green Revolution.

The Chicken

Every once in a while, my trip to the supermarket ends this way: Nearing the cashier I find myself standing at a display of barbecued chickens. Usually ten or eleven bucks each (fourteen these days), a whole chicken, legs trussed, cooked, and still warm. Sometimes I grab one with the hope that it might just enhance what's on the table for dinner.

These rotisserie chickens are staples of supermarkets, but even when we eat them we don't pay much attention to them. Sure, we reflect that the meat was just moist enough, or a little too dry, or what am I going to do with the rest of it (if there is a "rest of it"). But we don't *consider* it. This particular kind of bird . . . how did it end up in the supermarket?

Incidentally, it's actually remarkable that it's an entire bird, as that isn't the usual fate for chickens these days. The vast majority are carved up into breasts, thighs, and stir-fry chunks, the underlying assumption being that people today are either too busy to carve a chicken or simply lack the skill.

A Species on Fast-Forward

It used to be thought that the chicken is the heir to an eight-thousand-year lineage, but recent research has telescoped that time, so now it's believed the red junglefowl was domesticated around 3,500 years ago in Thailand, not at first as an edible species, but a revered one.* That leaves a long time for the chicken to change, but still the last half century or so has been extraordinary, and has, with much human help, broken world records. Those chickens raised for their meat, not their eggs (so-called broiler chickens), are now *four times* bigger than they were in the 1950s, and they reach that size much sooner than they used to. That's their past, but I doubt that's going to be their future.

That ancestor of your rotisserie chicken, the red junglefowl, is still alive and well. (Actually, maybe not "well" because they're interbreeding with barnyard chickens and diluting their gene pool.) You'd recognize this bird right away: the males look like barnyard roosters, the females like barnyard hens. They scratch around on the ground for food and only take to the air to find a roosting spot for the night. All of today's chickens go back to that bird, or mostly that bird with small genetic contributions from other junglefowl.

If there's any room for irony here it's this: It's possible that the red junglefowl was domesticated not to eat, but to fight. Cockfighting, where two males attack each other with their bills, and more dangerously, sometimes equipped with curved, pointed steel spurs on their legs, is illegal in North America, but it is still practiced in many countries, allowing the

---

* This same research suggests that because 3,500 years ago was about the time that rice and millet were beginning to be grown, perhaps junglefowl were attracted to human habitation by that food source.

claim that it is the world's longest-lived sport—there is a depiction of a contest on a mosaic in Pompeii.*

That the ancestral red junglefowl and the modern chicken are intimately related doesn't mean they're the same. The first few millennia of domestication passed with little effect on the size and weight of the ancestral version. But for the last several centuries humans have bred the bird to be bigger and meatier, and in the last few decades, well, the bird has been transformed.

Celebrating the "Chicken of Tomorrow"

The modern chicken was given a huge push in the late 1940s when the A&P grocery chain in the United States kicked off the "Chicken of Tomorrow" contest. Cash prizes were given to poultry farmers who could come up with a chicken that, according to one description, would be a "broad breasted bird with bigger drumsticks and plumper thighs." It worked: breeds of chickens with exactly that description were produced over the three years of the contest, putting the industry in a position to create a whole new, very large bird. There was even a parade held to culminate the project, complete with the crowning of the "Chicken of Tomorrow" queen.[3]

Here's how the "Chicken of Tomorrow" led to the chicken of today. The average red junglefowl female weighs about a kilo; males about a kilo and a half. That barbecued chicken I sometimes buy, a low-slung, widebody version of the junglefowl, probably weighs about a kilo, but it is definitely on the small side compared to those you'll find at the meat counter.

---

* It's arguable that fighting cocks might live longer than their cousins in the broiler cages; it's even been argued by Hal Herzog at Western Carolina University in his book *Some We Love, Some We Hate, Some We Eat: Why It's So Hard to Think Straight About Animals* that roosters raised for cockfighting live a life of luxury for two years, before they enter the ring and then, yes, there is a real possibility that they will die a violent death. But, he asks, is that a worse life than living six weeks crowded together with thousands of other chickens, guaranteed to be slaughtered? He thinks not. Some fighting cocks even retire scarred, but still alive. No broilers do.

As I said, the average chicken today weighs about four times as much as its counterpart back in 1957. If you want a deep dive into how this progressed, data from a side-by-side comparison of chickens from 1957 with chickens in 2005 tells the story in detail.

Both newborns started their lives weighing 40 or 45 grams (about an ounce and a half). After about two and a half months of constant feeding, the 1957 bird was ready for market. It weighed about 900 grams (two pounds) and could feed two. The 2005 version was ready for market faster, in less than two months; was more than 3,000 grams (six and a half pounds) heavier; and could feed six.

These days, a newly hatched chick can be converted into a 5-pound chicken in less than six weeks. The modern chicken's breast is 80 percent bigger than it was decades ago. Put simply and grotesquely, if human babies grew as fast as chickens, they would weigh more than 2,700 grams (600 pounds) at two months old.

Put just as simply but fantastically, if current growth improvements were to continue at the same rate for seventy-five more years, a newborn chick(en) would weigh 2 kilos (just under a pound), hatched from a 2.7-kilo (6-pound) egg, parented by a 160-kilo (350-pound) hen and a 180-kilo (just under 400-pound) rooster. Cockfighting would be the new UFC.[4]

Extraordinary weight gains, plus the ability to move chickens from hatching to market in weeks rather than months, has resulted in unheard-of numbers of chickens. But the demand for them has kept pace, if not forced it.

On average, each American consumes more than 45 kilos (100 pounds) of chicken every year, compared to 36 kilos (80 pounds) in Canada and 30 kilos (65 pounds) in Mexico. Add the growing consumption in China and the rest of Asia and, even with body mass rocketing as it has, this desire for chicken meat would have far outstripped production unless there were not just heavier chickens, but more of them.

How many are there? At any moment, an estimated 23 billion, egg layers and broilers together. Chickens weigh more than all the other birds in the world put together.[5] In fact, they probably represent the largest population of any bird in the history of the Earth.

For context, take the house sparrow. Honestly, it's hard to be in any city in the world and not see house sparrows, but there are only half a billion worldwide. Remember the fabled passenger pigeon? We hunted them to extinction, but descriptions of the size of their flocks, which literally darkened the skies as they flew overhead, suggest there might have been 3 to 5 billion. Let's say 5 billion at the most.

These numbers are trivial beside the 23 billion chickens, five times as many as there were fifty years ago. And given that the chickens raised for meat live less than two months—sometimes closer to one—then over the course of a year the number of chickens raised is much greater than 26 billion, likely closer to 70 billion.

Two thousand are killed every second.

In 1966, global production was 10 million tonnes or 11 million tons. By the mid-twenty-first century, production might exceed 181 million tonnes or 200 million tons, twenty times the levels in the mid-1960s. These numbers reflect one huge advantage chicken meat has over beef and pork: there are no cultural or religious proscriptions or taboos against it. But how on Earth can those enormous future numbers be achieved?

How chickens are housed, fed, and bred has created the conditions for these amazing weight gains. Human ingenuity at work. It's estimated that breeding for weight has been responsible for about 85 percent of the weight gain and more efficient feed the other 15 percent. Genetics prevails because there is such rapid turnover. It takes much longer to raise a generation of pigs or cattle. And, of course, today's genetic technologies hold much more promise.

Running Around Like a Chicken . . .

In 2004, the chicken was the first domesticated animal and the first bird to have its genome sequenced. That should open the door to even bigger, more rapidly maturing chickens for the dinner table and KFC, right? But simple math says that the two relentless trends of an expanding body and shrinking living space will surely hit a wall, and soon. That's not all: accompanying the massive weight gain has been a parallel increase in health issues, which are diverse but interconnected.

The issues center on the chicken's heart and bones. Rapid growth requires plenty of oxygen, but the expansion of the chicken's body, especially the breast, has outpaced the growth of the heart, leaving the heart incapable of adequately oxygenating the chicken's tissues. This leads to the accumulation of fluid in the bird's abdomen, sometimes triggering heart failure and sudden death. Up to 10 percent of all chickens with fluid buildup in the abdomen are discarded, adding an economic downside to what is already an animal welfare issue.

If deleterious changes were limited to the heart, fixing the problem might be somewhat straightforward, but the chicken's skeleton has been left behind as well. Faster growth brings a higher risk of bacterial infections of the leg bones, leading to decreased mobility and even fractures. Sometimes the leg bones are warped. Only a minority of modern chickens walk normally. While it's difficult to quantify pain, sometimes their walking behavior, like taking unusually short steps, is interpreted as discomfort.

Again, evidence suggests that several growth-promoting genes bring with them these higher risks of infections, lameness, and loss of mobility. (That mobility is a health issue is somewhat ironic given that in extreme cases individual chickens have a floor space equivalent to a sheet of 8.5 x 11-inch paper.)

These are both economic and animal welfare issues. That each of them has a genetic connection suggests that more targeted genetic techniques could alleviate some of the health problems while still maintaining an economically viable body size. Technologies like the vaunted CRISPR-Cas9 have enormous potential for precise gene editing, but changing even a single gene may have knock-on effects elsewhere in the genome, and the pressure to get the balance right is significant: right now chickens have been bred to generate about a kilo (2.2 pounds) of meat for every 2 kilos (4.4 pounds) of feed. Maintaining that kind of efficiency while breeding healthier chickens is the issue. More sophisticated genetics will not be the only answer: in the poultry literature it's easy to find arguments (and commercials) for subtly tweaking the mix of corn and soy in a chicken's diet to enable bringing slightly more chickens to market every year.

What Can Tech Do?

The future will likely bring not just genetic advances but an onslaught of new technologies to the poultry industry. Robots will be everywhere, mostly doing the mundane tasks of cleaning the floor of the chicken house and picking up eggs that have gone astray. But some extremely specialized robots are on the horizon.

Here is a description of one prototype: "The Gribbot consists of a transport system, a machine vision subsystem, a robot arm, and a compliant gripper."[6] When the Gribbot is finally deployed, it will remove a chicken breast from the carcass in two to three seconds, allowing it, according to the Norwegian company SINTEF, to replace thirty humans on an assembly line. This promises "huge savings for the industry and releasing the employees from highly repetitive tasks." (Or rendering them redundant.)

Carving the breast meat out of a chicken is in a different class from scooping ice cream—the Gribbot has to employ machine vision and mimic human dexterity seamlessly to remove the meat. Testing suggests it will do that, and more accurately and faster than humans, even though there are challenges: the meat is shiny and confuses the camera; it's also slippery and difficult to hold on to. But it's hard to be skeptical when there is already a robot that can debone an entire chicken.[7] This robot, developed at the Georgia Tech Research Institute, first scans the chicken to identify familiar landmarks, then uses that information to plan where to cut. The cutting needs to be as efficient as possible, avoiding bone but maximizing yield, and to do this the robot has to know the different "feel" of bones, tendon, muscle, and ligament. It, too, has yet to be deployed.

Both of these robots are examples of the value of speed and efficiency, but neither has anything to do with live chickens. Those technologies that do can benefit both the company bottom line and the welfare of the animals. An example on the verge of being operational listens to the sound inside the chicken house, the din produced by 25,000 chickens clucking, groaning, murmuring, expressing contentment or discomfort. Veteran poultry farmers understand that these sounds can be warning signals that the chicken house is too hot, or that the birds are having respiratory problems, but monitoring that many chickens is difficult. Technology will help but it first has to be able to extract meaningful signals from the general babble. Once in place, sound analysis would make life slightly better for the chickens and likely lead to fewer health issues for the farmer to be concerned with.

These technologies have to benefit the industry, and hopefully the chickens themselves. But in the rush to make the poultry industry hyper-efficient and productive, another concern has arisen: bird flu. Extensive crowding creates opportunities for bird viruses to spread quickly and mutate. In case we needed a reminder, COVID revealed just how commonly a mutation that enhances the infectivity of a virus can spread. There have already been bird flu scares, with brief flareups of human infections. There are eight types of bird flu on the go right now: in 2022 more than 50 million poultry died of bird flu in North America.* Viruses emerging from poultry barns can be transmitted to wild birds, which, especially during migration season, can spread disease quickly and widely. Obviously the fear is that one day a bird flu virus will jump to humans. There have been occasional such infections, but none has taken hold in humans with the ferocity needed for a pandemic—but it could happen.

Takeout

Chickens have been so changed by technology that, as one scientist put it, the bird is "shaped by, and unable to live without, intensive human intervention."

At some point, even though the poultry industry has taken steps to

---

* This number could eventually reach hundreds of millions.

improve the welfare of the chicken, it becomes difficult to think of it as a living bird and not just a highly industrialized way, a very efficient way, of converting feed, much of which humans wouldn't eat, into food that they will.

It's a variation on the theme that a chicken is just an egg's way of making another egg: it's also a way of making soy and corn into meat. The industry has relentlessly pushed efficiency and speed: the less food required to take the bird to market weight in the shortest time the better. But that can't continue. The health issues would mount, despite better genetic technologies. Public demand for sustainability and better living conditions will run counter to more meat from less feed.

What will likely happen is that the genome of the broiler chicken will be the target: genes from breeds that respond to some of these pressures will be introduced, and the modern chicken will depart even further, at least genetically, from the red junglefowl. The practical way to maintain meat production won't be to introduce entirely new varieties, like heat-resistant chickens that could thrive in the face of a warmer world, but to implant heat-resistant genes. Then the broiler chicken won't just be a fast-growing, short-lived automaton, but a fast-growing, short-lived, genetically fine-tuned automaton. Still available in rotisserie format.

PART III

Where Will We Live and How Will We Get There?

CHAPTER 9

Planes, Trains, and Something Elon Musk Said

In North America, you don't have to look far to see cities that were designed for cars. The land has been paved over with broad streets and vast parking lots and there is little by way of pedestrian and bicycle-friendly spaces. Once a city devotes itself to the car, it's hard to turn it around. In 2050, two-thirds of the world's population will be urbanites. The biggest cities will have the population that Canada and California each have now—close to 40 million people. What will those cities look like? What sort of balance of natural and human-made environments will they have?

A hint might come from the most-talked-about transportation technologies of the future, all of which emphasize one thing: speed. COVID-19 eliminated the home-to-office commute. But if offices move back to in-person attendance, then faster, more convenient travel options will have more appeal. But the most radical options are enormously expensive, and no matter what technology is considered, environmental impacts will play a dominant role. The success of any transportation technology will be determined by how it handles energy consumption and greenhouse gas emissions, in addition to the need for speed.

Transportation generates roughly a quarter of the world's greenhouse gases, so advances in ground transportation, like the self-driving car, may be a winner; China's 1,000-kilometer-per-hour (621-miles-per-hour) train might take maglev to the next level, and aviation's new turtle-and-hare combo—the supersonic passenger airplane and the blimp—may resurface!

Airborne

Aviation consumes gargantuan amounts of fossil fuels and emits corresponding volumes of CO_2. These emissions have doubled since the 1980s, now measuring more than a billion tonnes or 1.1 billion tons of carbon dioxide annually. However, because total emissions from all sources have grown at the same rate, aviation remains around 2.5 percent of the global total. While that figure doesn't sound catastrophic, if aviation emissions continue to rise while reductions are made elsewhere, it won't be good.[1]

Passenger air travel accounts for about 80 percent of aviation's total emissions, with the startling statistic that the 1 percent who are truly frequent fliers are responsible for more than half of the total emissions.[2] When it comes to the future, formerly bullish forecasts of 4 percent growth or more per year (or doubling between 2020 and 2050) have been whittled back to around 3 percent to take into account COVID-19 and its impact on travel. But even at that growth rate, in the absence of any reduction in emissions, by 2050 the aviation industry would be generating substantially more than double the emissions they do now.

There's the technology opportunity: major steps have to be taken to be more efficient in the air, thereby reducing fuel consumption and emissions at the same time. These might range from tweaking the design of the fuselage and wings, making them more aerodynamic, to running on jet fuel from recycled carbon dioxide. That's why it's surprising that we're being told that the return of the supersonic passenger aircraft is imminent. It's fuel efficient? It emits less greenhouse gas? Not so far. But it sure ticks off the increased-speed requirement!

The history of the supersonic aircraft was brief: The Concorde flew from 1969 to 2003.* After a crash in 2000, caused by running over debris on takeoff, passengers were reluctant to fly it. Air travel was still depressed after the terrorist attacks of September 11, 2001, and maintenance costs were rising. The main environmental impact of the Concorde was noise pollution. Because of sonic booms, flights were limited to crossing the oceans. Its sole attraction for passengers was speed, and it was pretty good at that: New York to Paris in three and a half hours, at a speed of just over 2,000 kilometers per hour (1,243 miles per hour).**

So consumer apathy, maintenance, and fuel costs, among others, killed the supersonic aircraft era. But now, with the Concorde still very much in people's minds, there's talk of a comeback.

An American company called Boom is ready to start testing a single-person, experimental supersonic plane called the XB-1. They hope to apply what's learned during test flights to construct a passenger aircraft called the Overture.*** While the XB-1 is using off-the-shelf engines, the Overture is supposed to fly on "sustainable" fuel.[3]

Which is the bigger challenge? Inventing a whole new engine, or

---

* The Russians had their own version, the Tupolev TU-144. It flew less often, crashed twice, and was taken out of service around the time of the Concorde.

** I don't know if any potential passengers were dissuaded by the fact that above 50,000 feet, where the Concorde flew, in case of an accident causing a breach in the cabin, depressurization would be too rapid for masks to drop down, so they would likely be unconscious within about ten seconds. The response instead was for the plane to descend rapidly to higher pressures; happily, the pilots' air supply would be maintained.

*** At the moment there is one other supersonic prototype being built by Lockheed Martin, but this one, the X-59, is not the forerunner for passenger flights. It is strictly a test of technologies designed to reduce the sonic boom.

coming up with the sustainable fuel Boom has promised? Which is the more important development for aviation generally?

The fuel.

The most sensible sustainable jet fuel is derived by processing food waste, vegetable oil, the products of genetically altered yeasts and algae, waste wood, alcohol from fermenting sugarcane or corn—there's a long list. All these sources can be converted one way or another into what's called "drop-in" jet fuel, one that can be substituted one-for-one for kerosene. Of all the potential sources, production is most advanced for fuels derived from cooking oils, animal fats, and vegetable oils. While tens of thousands of kilometers have been flown by aircraft using a mix of traditional and sustainable fuel, it still represents only a fraction of 1 percent of all aircraft fuel.

Yet there's progress: In December 2021, a United Airlines flight from Chicago to Washington, D.C., had one engine fueled with kerosene, the other entirely with sustainable fuel derived from sugar beets, corn, and agricultural waste. Previous demonstration flights, like those flown by the Canadian airline WestJet, used sustainable fuel blended with standard jet fuel.

You can tell from that example that we're nowhere near where we need to be with jet fuels. But there are other examples beyond sustainable fuel: one is hydrogen, and the first significant test flight of a partly hydrogen-powered plane took place in England in January 2023. The nineteen-seat plane built by the company ZeroAvia flew with one engine powered by traditional jet fuel and the other a 50-50 mix of electricity and hydrogen. The company predicts they'll have commercial flights in two years.[4]

Note the partial use of electricity to power that hydrogen-based flight. The issue with electricity-powered planes is that current batteries are just too heavy for the energy they yield—they're not "energy dense" enough. There are electric aircraft everywhere, but most are small and have only taken tentative flights. So far, we've had the test flight of "The Spirit of Innovation" built by Rolls-Royce (now the world's *fastest* all-electric plane). The "world's first fully-electric aircraft for *commercial* flight" has taken to the skies off Vancouver, and just a few months ago the Alice, a plane developed by Israeli company Eviation, made its first flight. The nine-passenger e-plane is dubbed the "world's first all-electric *commuter* aircraft."

Aspiration is not in short supply. One thing I can say for sure is that videos produced to advertise new electric planes have mastered the "art" of slow-motion openings, usually of hangar doors, accompanied by inspiring music. One thing that can't be said is how rapidly electric planes will be adopted.

Back to the Future

Now a hard left turn to contemplate the rebirth of the blimp. Yes, the promise of the blimp, or dirigible, seemed to be killed by the catastrophic crash of the *Hindenburg* in 1937.* Of course we've had the Goodyear Blimp appearing at Super Bowls and other outdoor events, powered by nonflammable helium rather than hydrogen, but beyond a few others hired to advertise, the skies aren't exactly filled with blimps.** Yet they, too, like the supersonics, might have a new future.

Blimps, or "airships" as they're called these days, are the polar opposite of supersonic passenger jets. They are very slow: 150 kph/93 mph might be the fastest you'll travel in one. Nor do they have great passenger capacity: one hundred would be close to a full load (of course, the bigger the airship, the greater the capacity). But passenger traffic isn't where airships shine—cargo is.

It's estimated that half the world's population has no easy access to a paved road, polar and tropical areas being prime examples. Imagine a place like that with no airstrip of any kind but a dire need for, well, you name it: supplies like water, emergency food or medical aid in case of

---

* It's worth watching: https://www.youtube.com/watch?v=CgWHbpMVQ1U. It's amazing that only thirty-five of the ninety-seven passengers onboard died.

** But they do have a toehold in pop culture. The Iron Maiden song "Empire of the Clouds," written by lead singer Bruce Dickinson—an airship lover—commemorates the *R101*, a British luxury airship that crashed and burned in 1930.

floods or fires, or infrastructure materials. Places where a plane can't land and a helicopter is too expensive. But a soft-landing airship is perfect—it fills the gap between fast and expensive, or as they say, "faster than an ocean liner and more cost-effective than an airplane."

Hybrid Air Vehicles in England hopes to be selling their airship, the Airlander 10, by 2026.[5] The company anticipates that besides cargo, there might be a market for short-hop passenger trips, like from Vancouver to Seattle or Belfast to Liverpool; slower than a plane for sure, but much faster than a ferry. The airship has the appeal of gliding at low altitudes, with very little noise, over the countryside.

One challenge for airships is to ensure they are more sustainable than the competition, whether air or ground or sea based. Electricity is likely the answer, at least in the future. Another challenge is the issue of what to do after offloading cargo. A lighter-than-air craft will start to lift off as its cargo is removed—definitely an inconvenience.

Hybrid Air Vehicles eliminates the problem with the Airlander 10; lifting it requires the engines to propel it to the point where aerodynamic forces combine with the helium's buoyancy. Lighter Than Air (LTA), an American company focused on humanitarian aid, is hoping to solve the problem by using hydrogen as fuel, the technology of handling it safely having evolved dramatically since the *Hindenburg*. Atmospheric oxygen combined with hydrogen from a fuel cell generates water, nine times as much water by weight as the original hydrogen, so an airship like this would be gaining weight as it flew, again lessening the chance of spontaneously rising as it's being unloaded. Aeroscraft, a California company, plans to cyclically compress (to be replaced by air) or release helium (to displace the air). Air being heavier than helium allows this system to keep the airship on the ground, or not. Three different approaches to the same problem might be an indicator of where the airship industry is at the moment.

Like so much of the speculation around future transport technologies, it's very difficult to say with confidence that airships will be a significant player in ten or twenty years, but there is a niche for them to occupy; whether it pays the bills or not is the question.

I suppose I should also mention suborbital or even orbital flights in a billionaire's rocket project. It's really pretty remarkable that even though there's been a lot of talk of space tourism over the years, today it is suddenly upon us. But like a few years ago, when the very rich ascended to the International Space Station with vows of bringing worldwide attention to the state of the Earth and our responsibility for it, the future of riding Virgin Galactic, SpaceX, or Blue Origin into space or close to it (rival billionaires like to debate exactly what constitutes "outer space") has nothing to do with sustainable transport on Earth. Exactly the opposite, in fact.

Suspended

It was not my goal to make multiple references to Elon Musk in this book, but here he is again. His "hyperloop" has gained much attention, even though it will likely never be built. Nonetheless, the saga of the hyperloop is a classic example of how not to believe every hyped tech idea. The hyperloop is designed to use the well-established technology of magnetic levitation (maglev), in which objects are suspended, held in midair by magnetic forces. The same technology underpins a related project, China's 1,000 kmh/621 mph train, which might actually get built.[6]

In 2012, Musk published his fifty-eight-page "Hyperloop Alpha" white paper.[7] He declared: "The Hyperloop (or something similar) is, in my opinion, the right solution for the specific case of high-traffic city pairs that are less than about 1,500 kilometers or 900 miles apart. Around that inflection point, I suspect that supersonic air travel ends up being faster and cheaper."[8]

Musk went on to point out that supersonic planes aren't useful for shorter distances, as the advantage of their speed is mostly realized in horizonal flight, and on short flights they would spend most of their time taking off or landing. Hence the hyperloop, which he touted as providing the speed of an airplane with the convenience of a train. Musk called it "a cross between a Concorde, a rail gun, and an air hockey table."

Here's how it's supposed to work: The hyperloop uses a combination of magnetic levitation and vacuum. Imagine a sealed, 500-kilometer-long (310-mile-long) tube, with a track and passenger pods inside. It should be as straight as possible to limit the centripetal force that comes into play when taking curves at high speed. The key is to suspend the pods in the air, using the same repelling forces that you can feel when you bring two magnets together. The pods have magnets on the bottom and the track is lined with electromagnetic coils. When power is applied, the coils repel the magnets on the pods and the pods rise centimeters up above the track. Friction having been eliminated, the only thing holding the pods back is air resistance, hence the establishment of at least a partial vacuum to remove most of that air. A complete vacuum would be too difficult to maintain, but an atmospheric pressure equivalent to an altitude of 180,000 to 200,000 feet (55,864 to 60,800 meters) would work almost as well. Then any forward push applied to the pods yields great speeds; the push is supplied by a variety of mechanisms.

What would it be like to ride a hyperloop? Seeing as how so few have had this experience (and then only at much lower speeds), you just have to imagine sitting with the twenty-seven other passengers, the pod closes, you feel a slight acceleration that's comparable to airplane takeoff, then you settle in for what is touted as being an incredibly smooth ride—after all, you're suspended in midair. And the view from the windows! Sorry, I mean the "interactive panels" or "augmented windows"—they're *not* windows; they provide you with images of the scenery outside, plus other trip information like where you are and when you'll arrive.

If you remove the tube enclosing the hyperloop, exposing the track and the pods, then the vacuum is lost and you essentially are left with a maglev train, still floating on the magnetic field between the train and the track, but now pushing against the atmosphere to move forward. But even these are pretty amazing. In 2016, a Japan Railway maglev briefly hit just over 600 kph/373 mph; China has put a train into service that should routinely hit that mark. (A reminder: that kind of speed is in the range of a jet aircraft.) In spite of the extremely high cost of construction, both countries are heavily invested in maglev trains. Japan hopes to have a Tokyo–Osaka line ready by 2037, which would reduce a three-hour trip to just over an hour. China, meanwhile, has announced the development of a maglev train that could hit 1,000 kph/621 mph, an absolutely unreal ground-based speed, as part of a program to establish several maglev lines in the most congested areas. This is no surprise: China has the world's largest high-speed railway network already, which connects to almost all cities with a population of more than a million people.

Timetables for all these projects were stalled by COVID-19, and some were even dramatically altered. Virgin Hyperloop (in 2016, Virgin Atlantic took over the company Hyperloop One) announced in early 2022 that it was letting go of half its staff, abandoning pursuit of a passenger hyperloop, and concentrating on freight. Shipping freight at extremely high speeds is safer and easier than dealing with passengers, and likely makes better business sense, but there is a lingering fear that abandoning the passenger part of the project will be a serious blow to the entire hyperloop industry.

It might not have gone anywhere anyway—there are lots of critics who doubted the venture from the beginning. It started in 2013 when the blog *Pedestrian Observations* went after Elon Musk's initial "Hyperloop Alpha" white paper, pointing out that among other things the forces passengers would be subjected to would ensure the hyperloop was "a barf ride." Some analysts argued that a hyperloop in Europe wouldn't have the capacity to absorb more than 5 percent of the commuter train travel there. There were fears that tectonically active California would make a San Francisco–Los Angeles route dicey because the track has to be extremely precisely laid and remain invulnerable to vibration or agitation. The installation costs are enormous compared to every other transportation system. Yet, as the proponents argued, once built, hyperloops offer relatively inexpensive and extremely rapid movement of at least freight. "Broadband for goods," it's been called, probably hopefully. Virgin's decision to stick to freight might give the hyperloop industry time to figure out whether passenger service would ever be worth it. Then journalist Paris Marx questioned Musk's motives for floating the idea.[9]

All is not completely lost, though. As Elon Musk himself pointed out, hyperloops on Mars would make perfect sense: the atmosphere there is only about 1 percent as dense as on Earth, so air resistance would be a nonissue.

Grounded

Self-driving cars are a new technology that seem guaranteed to take hold. Already cars and trucks, some semiautonomous, some completely so, have racked up hundreds of thousands of kilometers everywhere from American highways to small-town main drags. They are seen to be a game changer for more than one reason. First, a promised reduction of car accidents. Nearly forty thousand people die in car accidents in the United States every year; 1.35 million around the world. Distracted driving might be involved in nearly half of those, something that an autonomous vehicle is incapable of. Self-driving cars could dramatically reduce the number of motor vehicle accidents—there's little doubt of that.

But the details! What standards will self-driving cars be held to? Should they be twice as safe? Ten times as safe? Will consumers accept them quickly? Trust them?

The technology is only part of the equation—the way humans and self-driving cars interact is just as important.

A fatal accident in Tempe, Arizona, in 2018 made clear what self-driving uncertainty looks like.[10] A self-driving Uber Volvo (with an operator on board but not in control) was driving a circuit when at 9:58 p.m. it noticed a "vehicle" ahead—5.6 seconds ahead. It didn't alert the driver. Then it wasn't sure it was a vehicle. Then it was. Then, at 2.6 seconds, it identified the object as a "bicycle." At 1.5 seconds, "other." Then "bicycle" again. Finally, at 0.2 seconds before impact, the car alerted the driver, who grabbed the steering wheel at 0.02 seconds before impact, but it was too late. The car hit a pedestrian who was walking her bike. She died at the scene. Ensuing investigations were complex—and are still ongoing—but there was general agreement that momentary inattention on the part of the driver played a part in the accident.*

Of course, self-driving cars aren't supposed to require the constant

---

* The National Highway Traffic Safety Administration in the US has recorded 736 crashes and 17 fatalities involving Tesla's self-driving software since 2019.

attention of a human, at least when unleashed on the public. And there's no doubt that one day they will be and they will be up to the challenge of identifying a person walking her bike and taking the steps to avoid her. But for some observers that will be a hazardous time, when traffic is a mix of humans driving cars, self-driving cars with watchful—or not—humans present, and completely autonomous self-driving cars. This array of driving options is expected to last for an extended period of time. The human-only phase, while not all that safe, is at least familiar. The self-driving-only phase is expected to be very safe, because there will be no drunk drivers, no road rage, no speeders—no emotion. It's the in-between stage that threatens problems.

In a paper published in the *Proceedings of the National Academy of Sciences*, author Peter Hancock, a psychologist at the University of Central Florida, and his colleagues expressed concern over that mix of operators behind the wheel, and how that creates an imbalance between humans and algorithms.[11]

We may not always be good drivers, but at least most of us are aware of the intentions of others, even when we're semi-concealed inside a car. A significant part of the rules of the road—beyond traffic lights, stop signs, and posted speed limits—is each driver's assumptions about how others are going to behave. It might be too much to expect a self-driving car to understand the "courtesy" of letting someone slide into a stream of traffic even when they don't have the right of way. In addition, those rules of the road differ from city to city, even in the same country. Canadians know that driving in Montreal is different from driving in, well, almost anywhere else in the country. Los Angeles might qualify for that title in the US. Could self-driving cars acquire software updates to allow them to drive appropriately wherever they are?*

---

* Cruise LLC is an American company that has developed, by most accounts, the most advanced self-driving software around. They are already running a robo-taxi in San Francisco, having chosen that city specifically for its driving complexity. But their vehicles have been involved in sudden, unanticipated stops resulting in rear-end collisions, blocking traffic, and running over fire hoses at the site of a fire.

The authors suggest that because self-driving cars will adhere to a strict set of rules, they might even be bullied by aggressive human drivers. They conclude, "The dissonance between what the human knows of the driving world and what the machine is programmed to do will mean that during the approaching transition period conflicts between human drivers and autonomous vehicles are virtually inevitable."

A Car with Moral Constraints

Tailoring self-driving cars to their environment has even deeper connections to human behavior. A long-term, in-depth study called the Moral Machine comes to grips with some of the trickiest decisions a self-driving car might have to make. It's been called the largest experiment in moral psychology in history: millions of people from more than two hundred countries have had an opportunity to make the kinds of choices that a self-driving car might have to make.[12]

The Moral Machine is based on the venerable "trolley problem." The trolley problem comes in many forms, but the standard version has you driving a trolley that has lost its brakes. If you stay on the track you're on, you will likely kill five people who are standing on the track; if you switch to another track, you will kill only a single individual. What do you do?*

One school of thought argues that saving five is better than saving one, so switch; another points out that in the case of the five, you need do nothing, but saving them requires you to *decide* to kill someone.

The creators of the Moral Machine extended this dilemma to self-driving cars. They wanted to know what people in different countries would want their self-driving cars to do in similar, but more complex, situations.

The data revealed that people who played the online game would save a woman with a stroller over anyone else, followed by a girl, a boy, and a

---

* For the lighter side of the trolley problem, check out https://neal.fun/absurd-trolley-problems/.

pregnant woman. Next in line were doctors, athletes, and executives of both genders. Toward the end of the list were an old man, an old woman, a dog, a criminal, and finally, in last place, a cat. (Sorry, cat lovers.)

But the Moral Machine also gathered data on who was responding and where they lived. Western countries valued car passengers over pedestrians; Eastern cultures, including Japan and China, where there has always been a reverence for age, were not nearly as pro-youth. A southern group of countries had the strongest preference for saving women and people who are fit.

In many countries the preference was to save a pedestrian crossing the street legally versus one who was jaywalking, but in lower-income countries there was more tolerance toward the jaywalker. All this to establish that decisions made by a self-driving car in one country would be different from those in another.

There's obviously much more to consider in creating a self-driving car than simply more sophisticated arrays of sensors. The creators of the Moral Machine put it this way: "Never in the history of humanity have we allowed a machine to autonomously decide who should live and who should die, in a fraction of a second, without real-time supervision."

Yes, the Flying Car!

Is it even possible to write about the future of transportation without ruefully acknowledging the flying car? The poster child for failures of prediction, the comic book technology that the Jetsons enjoyed, the aircraft that populated the skies of *Blade Runner*, but never, you know, the real thing.

Amazingly, the flying car might finally be on its way.

Actually, the flying car has been here already. Cars and planes came into existence at roughly the same time in the early 1900s, and it wasn't long before someone had the bright idea of combining the two. In 1917, for instance, just a few years after the Wright brothers flew for the first time, American aviation pioneer Glenn Curtiss flew, or rather hopped, his "autoplane."[13] It was definitely weird. Imagine (or google) a dumpy little two-passenger car with a giant propeller in the back, a tail extending much farther back, and a pair of wings, biplane-style, on the roof. The contraption literally did hop, getting momentarily airborne, but it never really flew. It was definitely a hybrid car-airplane, but where could it be driven? A glance would tell you the wingspan was too large to make it possible to squeeze into any normal lane of traffic.

But cool ideas persist and evolve through generations of inventors. In 1937, a former associate of Curtiss, Waldo Waterman, married a Studebaker car engine with a single detachable wing and created the Waterman Aerobile.[14] The Aerobile was capable of about 110 kph/68 mph on the road, 170 kph/105 mph to 180 kph/112 mph in the air, had a range of about 500 kilometers or 310 miles, was cheap, and was apparently easy to fly—three Aerobiles attempted the flight from Santa Monica, California, to Cleveland, Ohio, a distance of more than 3,000 kilometers or 1,864 miles, and two actually made it. Alas, there was no demand. Only five were bought (by Studebaker) and that was the end of the flying car with the worst name.

The Aerobile was followed by the Aerocar in 1949.[15] Imagine the back half of a two-seater plane, complete with tail and wings, stuck on top of a small car. The Aerocar was designed to switch from plane to car to plane, each switch taking about five minutes. This one got approval from the US Civil Aeronautics Administration (predecessor of the Federal Aviation Administration); there apparently is still one Aerocar that flies.*

It's obvious from these few examples that while a flying car was a challenge that intrigued a very small number of entrepreneur-inventors, no one in the buying public was really interested. When you consider that the major advantage of a hypothetical flying car cited by consumers today is to reduce commuting time, it's not hard to see why. Road congestion wasn't the pressing problem decades ago that it is now.

---

* Chuck Berry's 1956 single "You Can't Catch Me" was inspired by the Aerocar.

Flying and Driving—the Odd Couple

It's true that many of the early flying car designs were unwieldy and sometimes not well thought out.* A "flying car" may sound easy to design and manufacture, but it's not. Planes and cars are designed differently because they do radically different things. Aircraft depend on lift, but cars need downforce to stay anchored to the road (hence wings on race cars). Aircraft fuselages are light and aerodynamic, while cars have to be solidly built to withstand collisions. A car's low center of gravity is not helpful aerodynamically, and even seemingly trivial features like a car's side-view mirrors create drag when flying.

The addition of the technology needed to fly, like wings and propeller blades, add weight, and in response the car has to add power. The wings can't be shrunk down too much; otherwise the vehicle won't get off the ground. Aircraft engines are too powerful for cars. And unless the wings can be folded to the sides of the chassis, or detached and towed, they make driving on actual roads problematic. Unless the propeller or rotors can be shielded, they're a safety hazard. Balance all of that if you can!

That's not a complete list but it should convince you that if a flying car were to be built, it probably wouldn't look like anything either you or the movies and TV have envisioned. It would likely be more accurate to imagine a hybrid between common low-cost drones and helicopters, the principal shared advantage being the ability to take off and land vertically.

In addition, you're not going to put a flying car in your garage. Even if you do, are you going to be able to back it out, look both ways for traffic, then accelerate down the road until you're airborne? While your neighbors are doing the same thing? On the other hand, imagine walking, biking, or driving to the closest landing site where you could board a two- or three-person flying car (probably with a pilot) and fly to your

---

* One attempt wed a Cessna Skymaster to a Ford Pinto, and while the aircraft was fine, this car has made the most lists of the all-time worst ever made. Sadly, the designer and the pilot were killed when it fell apart in flight.

destination. Possible? Maybe. And might that be faster than driving or taking public transit? Probably. Would it be cheap? Probably not.

Where do the assembly lines of flying cars stand right now? There are no such assembly lines yet, but there is much more interest and action than you might think. In 2019, *MIT Technology Review* magazine listed nineteen models of flying car either being designed or in prototype production.[16] Some have attracted funding from billionaires; some are being developed for Boeing, Airbus, Toyota, Hyundai, and Uber. So far, only two are fully capable of both highway driving and flight; most are drone-derived, including quadcopters, a single-rotor gyrocopter, and a hexacopter. There are also multiple versions of vertical takeoff and landing (VTOL) vehicles.

Who is going to own one? Not many, apparently. Uber's interest is prominent among the companies because they are promoting the idea of ride-sharing: Uber, despite having sold off its flying car project, apparently continues to believe in it, investing in the company that bought it.[17]

Hyundai's aim is to build and integrate a chain of services: use your app to ride a scooter to the vertical landing site; catch an electric VTOL vehicle to a landing site away from downtown; then ride-share (or bike) home.

These companies are making significant financial bets. Counting them and all the others, it's estimated that close to a billion dollars has been invested in what some call the "air mobility" industry. It's a few years before any air taxi system is likely to be up and running, and the research sucks in hundreds of millions of dollars a year.

Still the Odd Couple

One of the biggest challenges is the power supply. Internal combustion engines are a nonstarter for these vehicles of the future: those engines are very loud, too loud to be all over the sky in the city. Adding hundreds of engines' worth of carbon dioxide would also be a huge disincentive.

But while everyone talks about using electric power, and there are indeed planes that have accomplished it—it's not impossible—batteries are obviously up to the challenge of moving a ground-based machine, but a flying one is different. The power consumption is much higher and there's also the issue of weight. The battery must be light (it's powering a plane) yet it must also be powerful (it's traveling significant distances). Charging times at the moment would be inadequate for a busy airborne taxi service. Unfortunately, electricity isn't as "energy-dense" as jet fuel, so the challenge grows. Jim Heidmann, manager of advanced air transport technology at NASA's Glenn Research Center, was quoted in the *Guardian* as saying that with today's batteries, powering a Boeing 737 would require a battery as big as the plane itself. Perhaps unnecessarily, he added, "That's just not feasible."[18]

If the battery issue can be resolved the environmental impact would be greatly reduced. One study concluded that on a projected 100-kilometer or 62-mile trip, a single-occupant flying car (a "taxi") of this type would emit fewer emissions per passenger than a single-occupant car with an internal combustion engine, but more than a similar battery-operated car. But add three passengers to the taxi and it beats both alternatives. The issue revolves around the number of passengers and the length of the trip. Would passengers be more willing to share a ride given the anticipated speed of the journey? The researchers pointed out that VTOL vehicles use most of their energy taking off and reaching altitude, so the longer the trip, the more efficient it is. Of course, this puts even more pressure on the battery, underlining the importance of better battery technology; without it, the full-scale launch of these twenty-first-century versions of flying cars will be significantly delayed. But we're used to that!

Batteries are not the only critical issue. Who's going to fly them? Some of the proposed versions will require a pilot; all will require training and licencing. (Getting your VTOL license?) So maybe someone with enough money to buy a vehicle and hire a pilot? But maybe you really wouldn't need to own one, as long as there's a system of some kind in place, designated areas for taxis/planes/cars taking off and landing vertically. To take full advantage of being able to travel above the congestion, every step of the way has to be convenient. The one overriding advantage that flying cars are supposed to have is dramatically reduced commuter time—a thirty-five-minute trip turned into a seven-minute short hop. There's no point saving thirty-five minutes on a trip and then somehow wasting it on the ground.

Fantasy? Reality? Comedy?

In an urban setting you can imagine designated air corridors, like aerial bike lanes, accompanied by "sky harbors." The third dimension opened up, almost the way it's been portrayed in science fiction movies, but with technology that now seems within reach. It can't be a free-for-all if the air begins to reach saturation with drones, single-person jet cars, air taxis, single-engine planes, airliners, and birds. Modern cars are loaded with sensors to put them on the verge of being reliably self-driving. Flying cars living in a three-dimensional world will need many more, likely linked to a super-advanced AI. Ideally, from the energy point of view, you'd want as few takeoffs and landings as possible, because horizontal flight is so much more efficient. The long and more straightforward rural routes should then prevail, at least at first, so flying cars are less likely to establish in cities.

Still, in the world of e-commerce, ride-sharing, and aviation, there's a buzz about flying cars. Financial institutions have pegged the future value of the industry at anywhere from $800 million to more than $1 trillion in the late 2030s or early 2040s, although some of those projections were made before the delays created by COVID-19.

But there is enough uncertainty about the flying car's future (even after all that science fiction!) that diametrically opposed views of their value still exist. In 2018, the Center for American Progress issued a report arguing that if the industry fails to thrive, the impact of a few uber-rich having access to a flying car wouldn't matter much; it would be "just another toy in the stable."[19] The article contends that it would be more likely that flying car trips would now be *added* to car trips, rendering the comparison between the two irrelevant, leaving social and environmental downsides with little or no accompanying benefit.

Some will be swayed by those arguments, but community reaction will be important, too: Are you okay with your neighbor flying low over your backyard, close enough that his "low noise" engines are deafening? (Excess noise from propellers or rotors is such a concern that any contending designer incorporates several features specifically to minimize noise.) Remember looking up into the clear blue skies of summer before there were low-flying flying cars? Are you content that they're draining a wetland to install a landing site?

But let's not be negative. Perhaps it's no surprise that the director of the NASA Aeronautics Research Institute, Parimal Kopardekar, sees a different, brighter future. He envisions city design changing to accommodate rooftop landing and intra-urban flyways, freeing up ground-level green space and decreasing urban congestion—all this by 2045.[20]

I think everyone is so focused on "shortening the commute" as the raison d'être for flying cars, they're missing some of the future possibilities. So cast your mind forward to a time when, at least according to a recent BBC report, "Owning a VTOL (Vertical Take-Off and Landing) could become as affordable and ubiquitous as owning a bicycle."[21] I can't imagine that literally being true, but what if single-passenger flying cars did reach a point where they'd cost about as much as, say, a high-end sports car today? That could open the door to a huge tourist adventure/fitness market. You could hang out in previously inaccessible places, have a special "off-road" version of a flying car, or you could have a race like the Sahara Rally, flying cars hopscotching across the country, on-road, off-road, and in the air.

Anyway, it's getting too *Jetsons*-ish here, so I'll close on that sporty note.

CHAPTER 10

Where Will We Live?

Striking, isn't it, that science fiction depictions of future cities present gloomy scenes of clusters of immense high-rise buildings, punctuated by ground-based and aerial transportation routes, with dark, dangerous, and crowded spaces on the ground below. The original *Blade Runner* film of 1982 is a great example. With the oncoming crush of urbanites (6 to 7 billion by 2050), can we avoid such desperate cities? What is a Smart City? One that somehow manages to provide healthy living to all its inhabitants, that has room to walk and bike; is fully wired, secure, and entertaining. Will cities a quarter of a century from now also incorporate natural settings, known to improve human health and well-being? The dystopian visions of future cities have no trace of nature (with the possible exception of rats); could we replace dystopia with biophilia?

In 2017, Sidewalk Labs, the technology-oriented urban design subsidiary of Google, created a stir in Toronto when they were chosen to design a "smart" community on the city's waterfront.* It would be called "Quayside."[1]

---

* Sidewalk Labs is self-described as "a testbed for emerging technologies, materials, and processes."

They created a second stir in 2022 when they abandoned the project, although by then the seeds of doubt had been widely scattered. Sidewalk blamed disruptions caused by COVID-19 for their decision to pull out, but there had been complaints about their plan from the beginning.

The original idea was data driven, with a plan to build a community "from the Internet up." When fully developed, the twelve acres of land close to downtown Toronto would have self-driving taxis, heated sidewalks, garbage collection without garbage collectors, and beachfront office towers made of wood. Mass amounts of data would be collected— "urban data," as Sidewalk called it—gathered largely from public spaces. They were well aware of privacy concerns and assured data security, but in the end they were unable to convince the panel assembled to evaluate their plan. "Tech for tech's sake," a comment from the panel, is as memorable as "from the Internet up." The general feeling was that Sidewalk was too focused on the tech and not enough on the people.

The Sidewalk plan has now been replaced. An international team has proposed a development—also called Quayside—but with a lot less silicon, more green space, an urban farm, and affordable housing. Journalist Karrie Jacobs, describing the new Toronto plan in *MIT Technology Review*, was struck by the amount of greenery: "The renderings are so loaded with trees that they suggest foliage is a new form of architectural ornament."[2]

COVID-19 might have given Sidewalk the opportunity/excuse to make the decision to walk away without seeming to bow to the immediate and sustained criticism of the project. Regardless, does this example say anything about the practicability of so-called smart cities? Some of the proposed technology was seen by the panel as simply being unnecessary—the commonly voiced perception on the Toronto side that Sidewalk was arrogant didn't help. On the other hand, ideas like improving public transit and reducing private automobile use to 10 percent seem like good ideas. At the very least it argues that no matter how excellent the technology, local circumstance and input play a crucial role.

Smart Cities

Smart can be something as simple as well-planned public transit. The city of Curitiba, Brazil, is acknowledged to have one of the best: buses run as frequently as every ninety seconds, stations are comfortable, and this aboveground system runs as smoothly as a subway. Seventy percent of commuters in Curitiba use public transit.[3]

However, most definitions of smart cities (and there are lots!) emphasize digital technologies, as did Sidewalk Labs. The city of Chattanooga, Tennessee, together with Oak Ridge National Laboratory and the National Renewable Energy Laboratory, has built a "digital twin," a detailed 3-D model of the city that is up-to-the-second. Data from traffic cameras, 911 calls, weather events, and hundreds of other sources feed into the digital twin and can prompt a response, like tweaking the timing of traffic lights, in the real twin.[4]

Seoul, South Korea, has vowed to be a "capital of big data." They're deploying sensors to monitor traffic, installing thousands of surveillance cameras (with the obligatory claim that they are only to prevent crime), and providing free Wi-Fi on all city buses. But these technological additions, even when appropriate and helpful, are never enough in and of themselves. And there are the ever-present risks: if a technology exists, it will be deployed whether it's needed or not, and surveillance devices of any kind must somehow be surveilled themselves.

Nonetheless, enhanced technology will be ubiquitous in how cities are shaped as we approach 2050 and its accompanying urban crush. But what else? Somewhat lost in our love affair with technology is the idea that nature should also be an important ingredient in the creation of the smartest cities. Actually, nature can contribute in two quite different ways. Biomimicry is one, building a city using nature as the model. The second is connecting nature to the city's inhabitants. The first can improve the immediate environment; the second provides immediate benefits to the inhabitants.

Imitating Nature

As biomimicry expert Janine Benyus says: "We're not the first ones . . . other organisms, the rest of the natural world, are doing things very similar to what we need to do, but in fact they're doing it in a way that has allowed them to live gracefully on this planet for billions of years."[5]

As an example, China is designing or refurbishing cities to become "sponge cities," getting rid of hard, impermeable concrete surfaces that channel water away and replacing them with natural or engineered terrain (permeable surfaces, wetlands, vegetation, and creeks) through which water can move slowly, allowing time for it to soak into the earth rather than rushing away. Rainwater is degraded the moment it enters the sewage system, but in a sponge city, most of it doesn't get that far. Pittsburgh, Pennsylvania, is doing a sponge retrofit by using concrete blocks filled with crushed rock, rather than unbroken concrete. Water sinks through the bricks rather than washing across the surface.[6] In this case, as in many others, the principal goal of holding on to water that otherwise would drain away uselessly has additional, perhaps unexpected, benefits. A garden that collects rainwater and is populated with native plants provides a new environment for pollinators, and retaining groundwater is a hedge against the development of the particular variety of sinkhole called land subsidence.

But as you can tell from that last paragraph, biomimicry isn't always that thrilling, or even engaging. It needs a story—and it has one. It is about the Shinkansen, Japan's bullet train, and its kingfisher-inspired redesign.[7] In 1989, Eiji Nakatsu was given the job of redesigning the Shinkansen, both to solve a problem *and* improve the train. He also happened to be an avid bird-watcher—but never mind. The problem was a weird situation created by the blunt design for the nose of the Shinkansen. As soon as it entered a tunnel, the train shoved air ahead of it and gradually accumulated a mass of air that held it back, until the moment the train emerged from the tunnel. Then the mass of air exploded from the mouth of the tunnel, creating a shock wave that sounded like a sonic boom. The faster the train, the louder the shock wave. It has a name: tunnel boom (as opposed to sonic). It's pretty loud: people 400 meters or 0.24 miles away found it disturbing.

Nakatsu wanted the Shinkansen to be able to slice its way through the air inside the tunnel to minimize the shock wave—and the sound—when it came out the other end. He knew that the common kingfisher can dive straight into the water almost without disturbing it. Its pointed, tapering bill cuts through the water, instead of pushing it. It's pretty impressive given that when they dive they're flying at top speed, leaving one medium and instantly entering another that's hundreds of times denser.

When the newly redesigned kingfisher-like Shinkansen finally hit the track, it was 10 percent faster and used 15 percent less power. Why? Mostly because there was 30 percent less air resistance. A neat example of translating the bio into the techno and enhancing the life of residents close to the Shinkansen tracks.[8]

But such translations can be misleading. My favorite example is the Eastgate Centre in Harare, Zimbabwe.[9] The Eastgate Centre was built to maximize air circulation. Architect Mick Pearce designed the structure to circulate air in roughly the same way a termite nest of the same size would. It's called "passive internal airflow," a design that takes advantage of local wind patterns and temperature differences to move air naturally through a structure. They didn't even need to install air-conditioning, the building was that much cooler, allowing them to charge less for rent. The Eastgate also uses 15 percent less electricity than before.

But there's a twist: that idea that passive internal airflow is what keeps the termite nest environment livable? That it's all about air circulation? It's been discarded and replaced by the idea that the nest "breathes" like a lung, warm air being drawn in during the day, then pulling a U-turn at night as temperatures fall, hauling the day's buildup of carbon dioxide out of the nest with it.

So the principles upon which Eastgate was designed were based on a false understanding, too—but it works anyway! Somewhere in the push to drive better air circulation they stumbled on some inadvertent benefit. Nothing wrong with that. And maybe in the long run each of Eastgate's features will be shown to have environmental benefits, however small.

Designing by mimicking nature will be one part of the future city; if we're smart there is another, probably more important part.

The Need for Nature

Many studies have established that humans are attracted to landscape images of a watery, treed landscape extending out to the horizon from an elevated point of view, where there is some evidence of human habitation in the foreground, and hills or mountains in the distance concealing what lies beyond. A certain amount of complexity, but not too much, and an element of seeing without being seen (the elevated point of view). There have been discussions about whether this might be an innate human preference dating back hundreds of thousands of years or a more recent cultural phenomenon, but it definitely exists.

If that is the view we like the most, we're likely to be frustrated. Most cities are hard-pressed to furnish this kind of perspective, with the exception of city parks, gardens, and landscaping (and maybe the view from the penthouse), and those that do almost always limit it to the outskirts. Today, half the world's population of 8 billion people live in cities; that number is predicted to increase to two-thirds (of 9.8 billion) by 2050. You could conclude that in the future a majority of urban dwellers will not look out their windows on the kind of scene they would prefer. How important is that?

It's not just the view: there is an abundance of research on the impact of nature on the health and well-being of humans, and while the conclusion is settled, the details are still being worked out. In 1981, Roger Ulrich, a professor of architecture and landscape architecture, published a paper with the title "Natural versus Urban Scenes: Some Psychophysiological Effects."[10] Ulrich had already established that natural scenes, but not urban ones, improved the "emotional states" of people who were stressed. He extended the idea to people who were deemed to be unstressed, hoping to establish that scenes of nature had far-reaching benefits. He presented sets of slides, 240 in all, to volunteers. They represented three different kinds of scenes—nature with water, nature with vegetation, or urban with neither. None was particularly spectacular; none had images of humans or animals in them. Volunteers judged how information-rich and pleasant to view the slides were. In a second step, rather than relying solely on the participants' testimony, he measured their heart rates and brain alpha wave activity as well. An accelerated heart rate was interpreted as a rise in stress, as was a decline in alpha wave activity.

In the first part of the experiment, the natural scenes were judged to be much more pleasant than the urban ones; the difference was especially dramatic with those slides depicting water. In fact, only six of the water scenes rated lower in pleasantness than the *most* pleasurable urban scenes. All three scenes were comparable in terms of their amount of information or complexity.

In the second part of the experiment, personal testimony, brain waves, and heart rate were used to test the state of relaxation or stress the volunteers experienced when looking at the slides. Again, slides with natural landscapes, especially water, were the most "wakefully relaxed," as Ulrich put it. He pointed out that these effects weren't across the board; natural images were most effective at reducing sadness and fear. Ulrich's conclusion: "People benefit most from visual contact with nature . . . when they are in states of high arousal and anxiety" (which might be a good characterization of urban life in 2050).

These experiments are now several decades old, but the conclusions are rock-solid. A recent survey of twenty thousand people in England established that two hours spent in nature a week was associated with better health and feelings of well-being.[11] To dispel the suspicion that those who can get out and enjoy nature are healthier to begin with, a follow-up study looking at people who had recently moved showed that those who lived in greener spaces were happier (although the effects aren't limited to happiness, but also include improved memory and attention). Ulrich's work tied back to the landscape preferences I just mentioned. If, as some have argued, they stem from our *Homo sapiens* ancestors on the African savanna, they make sense: standing where you can see all around you allows you to be able to see danger approaching and know where the escape routes are. Fear of enclosed spaces is the opposite, and many intense urban scenes are exactly that: high-rise buildings encroaching on the sidewalk, hidden alleys, and hiding places.

More recently, Ulrich's discovery that water was even more beneficial than greenery has also borne out. A cell phone tracking experiment showed that when people were closer to water they reported a significant boost in mood. While there are other social factors that weigh more heavily on feelings of well-being, like employment and relationships, the water effect seems to apply regardless of income level.[12] You can see the same effect operating in a science center that has a simple water table equipped with plastic boats and waterwheels—children will play there forever. There's a reason people like Amsterdam and Venice!

How cool is it when you bring biomimicry and the psychology of nature together: Courtyards are a common architectural feature in Copenhagen, Denmark. Straussvej, its "courtyard of the future," features rainwater from roofs and pavement flowing into a central pond (the system can handle the proverbial hundred-year rainfall). The pond is surrounded by a garden, so that the design is full biomimicry.[13]

It would help to better understand the healthful effects of nature, to identify what exactly "nature" is and how it wields its influence. First, the dichotomy between urban and natural isn't as easy to draw as you might think. Asking people to view an underground parking garage versus a forest isn't really that helpful. Who feels totally comfortable in an underground garage? In the same way, a soccer field, as green as it is, might be more restorative than the parking lot, but that isn't that informative about nature/urban differences. Even urban streets where buildings are not menacingly tall and close to the sidewalk, where the sidewalks themselves are wide and tree-lined, where vegetation is not monocultural but diverse, have beneficial effects on those experiencing them.

Experiments in Japan, the land of "forest bathing" or *shinrin-yoku*, revealed there is more to the beneficial effects of being in a forest than just seeing it. Sitting in a chair looking out at a forest is great: blood flow to the brain increases and mood heightens. But sitting in the same chair, still surrounded by the forest but prevented from seeing it by a screen, also lowers blood pressure and stabilizes heart rate.[14]

More Japanese research showed that spending time in the forest enhances activity of the immune system. Both nurses and middle-aged businessmen, having spent three days in the forest, exhibited higher numbers of natural killer cells and boosts in their activity. Natural killer cells are important sentinels/soldiers in the immune system. The boost in natural killer cell numbers from three days in the forest was still elevated after a month.

Having detected significant amounts of airborne plant-based volatile organic compounds in the forest experiment, experiment designer Qing Li, a doctor at Japan's Nippon Medical School, then modified the approach by exposing volunteers to those same chemicals but indoors, in a hotel room, and again found that natural killer cell activity was heightened, although this time there was no apparent boost to their numbers. Still, apparently the beneficial effects of forest exposure are not the only emotional impact of nature.[15]

The Call of the Wild

The positive influences of the forest—boosting the immune system, increasing feelings of well-being, lowering blood pressure—are so subtle that they may go unnoticed at the time, but they are potent. There is a completely different set of positive emotions experienced when immersed in nature that are anything but unnoticed: the joy, thrill, awe, and occasional jolt of fear encountering wildlife *in the wild* rather than at a zoo.[16]

I hardly need to provide evidence, but here's some: There are fifty million bird-watchers in North America alone (although as evolutionary biologist David Barash points out, the hobby is generally doing better than the birds themselves), nature artist Robert Bateman has created more than seven hundred paintings, whale watching is popular around the world, and highways in national parks are clogged with cars full of people rubbernecking at an elk. In Yellowstone National Park people trying to get a nice close-up video of a bison get tossed by the subject. Forty thousand years ago, our ancestors were painting animals on cave walls, and there's no reason to think they were the first to be so fascinated.

We're mad about wildlife and should be doing everything we can to ensure it persists.

An emotional connection is not the only reason. They also protect our health. A recent study in the *Proceedings of the National Academy of Sciences* showed that living in proximity to a rich biodiversity, a wide range of species, seems to dampen the transmission of animal diseases to humans, the so-called zoonotic diseases like AIDS, Ebola, and (likely) COVID.[17] This study revealed that areas where human activity and settlement have encroached on and reduced the richness of wildlife are the same areas where animals most likely to transmit disease to humans have the easiest access to us. Exactly why biodiversity protects us isn't yet clear. You'd think the more species nearby, the more pathogens and the greater the chance that one will jump the species barrier to us. But it's apparently not so, and just as well: those disease-carrying animals that take advantage of low biodiversity are usually nonnative species. More of them means a further reduction of the "naturalness" of our living spaces. These so-called invasive species are doing exceptionally well in urban settings, but also in the wild or wherever they end up. Some just seem to have the necessary adaptations to thrive when transported by humans—deliberately or inadvertently—to ecosystems they've never before experienced. They are a diverse set: cane toads in Australia, Pablo Escobar's hippos in Colombia, Burmese pythons in the Everglades, and starlings everywhere in North America.* But this is nothing to celebrate! In 2016 a team of Australian scientists established that invasive predators have played a role in eighty-seven bird, forty-five mammal, and ten reptile species extinctions. Those numbers were likely underestimates because at the time there were an additional twenty-three critically endangered species that were judged to be "possibly extinct."[18]

The rat is a perfect example of an invasive species, an amazingly

---

* Drug lord Pablo Escobar had established a personal zoo at his headquarters in Colombia. When he was killed, most of the animals were sold off, but four hippos, a male and three females, were set free. That was the early 1990s. Today there are an estimated two hundred hippos wandering the watercourses in that part of Colombia.

successful animal at adapting to hitherto unexploited territories. The only continent that has no rats is Antarctica.* The consequence of their arrival is almost always the decline of a native species. A comprehensive survey of extinctions over the last several centuries revealed that rats—honorable mention to cats—were responsible for more extinctions than any other mammal (although to be fair to cat lovers, dogs weren't that far behind).** The dodo is a perfect example, snuffed out in a few decades by ships' rats, sailors' dogs, and sailors.

I guess you can admire the adaptability of invasive species, but not their impact. But the destruction of wildlife by invasive species (acting more or less at our behest), as bad as it is, pales in comparison to the direct impact we have ourselves. We don't really need help: birds, animals, and insects are disappearing at an ungodly rate. You don't have to look hard to see why.

Here are the numbers: Almost half the world's bird species are in decline and one in eight is in danger of extinction. The World Wildlife Fund's *Living Planet Report 2022* concludes that 69 percent of the world's vertebrates (fish, amphibians, reptiles, birds, and mammals) have disappeared since 1970.[19] Imagine that.

At the same time, and related, the so-called insect apocalypse is in full flight. Some estimates claim that 75 percent of the world's flying insects have disappeared over the last three decades. Imagine that, too.[20]

The numbers are nearly incomprehensible, but the causes can be identified. Number one, without a doubt, is climate change. Ecological systems tuned by millions of years of evolution can be disrupted in much less time than it takes to establish new ones. Some species can adapt; many cannot. Taking into account hunting and fishing at unsustainable

---

* The Canadian province of Alberta, where I live, claims there are no rats anywhere in the province. That's a landmass of about 660,000 square kilometers or 372,822 square miles.

** Let's not forget that while studies vary in their estimates, *at a minimum* cats in North America kill at least a billion birds every year.

levels, overuse of herbicides and pesticides, *and* pollution on land and in the oceans, you can only marvel at the tenaciousness of those species that survive.

Putting these threads together makes it clear that we aren't winning the battle to preserve nature, whether for altruistic reasons or even to better our own collective health. And more and more humans are mobbing into cities and away from nature. What can be done? Are there technologies that can help sustain wildlife?

Of course, there are technologies that will be applied to increase sustainable hunting, fishing, agriculture, and energy generation. Changes are under way in all these areas already. But the risk here is that we have an immensely strong bond with wildlife, a bond that sometimes overwhelms good sense. For instance, we have an unfortunate history of wanting to mess with wildlife. Besides the obvious hunting, we're also fond of redistributing animals, moving species around like chess pieces. Usually there's a "good reason." Like bringing all the birds referenced in Shakespeare to North America (artsy), or introducing cane toads to Australia to prey on sugarcane pests (science-y).

Sometimes it even works: the successful reintroduction of wolves into Yellowstone and the resulting dramatic changes in the ecology (for the better) is one example.[21] Here's a more controversial one: How about reintroducing the cougar to eastern North America? We drove them out; we can bring them back. Yes, if you live in their neighborhood it's important to be aware that they're around, you have to be careful where and when you hike, and they can be hazardous to livestock but they play an important role as a top predator. That can actually be worth money: If cougars were introduced to the eastern United States and began thinning the herds of deer, collisions between cars and deer would be significantly reduced. Four hundred and fifty-eight Americans are killed every year in confrontations with wildlife: 440 of those are car-deer collisions! In 2016 an American team of wildlife scientists estimated that 155 deaths (human) and 21,400 human injuries could be prevented by introducing cougars, and more than $2 billion saved, over three decades.[22] And they weren't taking into account the more recent research on disease protection afforded by biodiversity. Who knows whether ultimately economic/scientific arguments will prevail in that case.

Billions are spent worldwide on conservation. There are some triumphs like the increasing populations of the Eurasian beaver and the European bison, but it's hard to get excited about those examples when insects have declined so drastically that you can go for a long highway drive and not have to clean the dead insects off the windshield. Of course, seeing a herd of bison in the wild is thrilling; seeing a swarm of midges isn't.

Are we already in the sixth mass extinction? There's disagreement over that claim, partly because it's difficult to tally the actual number of species dying out. The definition of a mass extinction is that three-quarters of all species die out in a "geologically short time," which, being geology, could be two million years. But it could be much less, and our current rate of species depletion is at least competitive with the earlier mass extinctions. This has triggered some radical thinking that combines technology, our powerful bond with wildlife, and our urge to manipulate it. It's an approach that leans heavily on a vision of the future that captures our romantic views of "the way it used to be." It's compelling, but it very likely won't work and might, at the same time, harm other conservation efforts. It's called "de-extinction."

Bring 'Em Back Alive

Rapid advances in genetics like CRISPR and the ability to recover ancient DNA have made it possible to actually contemplate bringing back species that no longer exist. If you have enough preserved DNA of an extinct animal that you can assemble its entire set of genes, its genome, in a sense you have the animal in your hands. Many steps remain, but at least there's a path to re-creating the animal. There are experimental and huge technical hurdles in the way, but the *idea* is not impossible. But does it even make sense to try?

Setting aside sophisticated genetics for a moment, the first serious attempt to bring an animal back alive was about a hundred years ago. The German Heck brothers were commissioned by Adolf Hitler to re-create the aurochs, the huge two-thousand-pound ancestor of all modern taurine cattle (the Indian aurochs gave rise to the Zebu cattle). The last aurochs died in 1627 in Poland. The Hecks tried to breed cattle all the way back to the aurochs itself, the wild ox, retracing as best they could the steps that led to modern cattle, at least back to the early 1600s. Choosing which cattle to use depends on the features they share with their ancestor: Does this modern cow have the right coat color, the aurochs-like horn, the desired temperament (bad)? But they never really got a modern aurochs—so much for Hitler's dream of an animal symbol of such unusual size and strength and with a temper to match.[23]

More recently, *Jurassic Park* has firmly planted the idea of bringing back extinct species in the public mind, and that included some scientists who were inspired by the movie to dedicate their careers to de-extinction science. On the other hand, no one is thinking of bringing back dinosaurs. It's going to be hard enough to bring *anything* back, even if it became extinct much more recently than *Tyrannosaurus rex*, like the thylacine, the Tasmanian tiger—there's actually a film clip of this animal online. It might have finally died out in the 1960s somewhere in Tasmania. Still, bringing it back would be neither easy nor quick.

It's not well-known that one animal has been brought back from extinction: the Pyrenean ibex.[24] This animal was once common along the Spanish-French border but declined rapidly, and by 1910 there were only forty left. The last animal, a female, died when a tree fell on her in January 2000. Spanish researchers recovered some cells from skin samples that had been taken from the animal earlier, produced clones from the cells, implanted more than a hundred embryos, and one reached term. At first the newborn female seemed fine, but she died a few minutes later of lung abnormalities. Nonetheless, she was the first animal to be brought back from extinction. Unfortunately, her early death meant the Pyrenean ibex is also the only animal ever to have gone extinct twice.

Why is it tempting to think of bringing back any creature? There would definitely be nerdy interest in confirming that living dodos really were that plump, or Carolina parakeets that gaudy. And the southern gastric-brooding frog? Fabulous! But that's specialist stuff and not nearly gripping enough to launch an expensive project to bring it back to life. On the other hand, the woolly mammoth, the passenger pigeon, definitely are.

Rephrasing the question, why would anyone want to spend the money necessary to bring back even a species as fabulous as the woolly mammoth or the passenger pigeon? There's no doubt in my mind that underlying the efforts to de-extinct a species is the romantic notion of what it must have been to see them in the past—a museum diorama brought to life. How amazing would it be to visit a park in the Northwest Territories where herds of mammoths roam? A giant long-haired, cold-weather elephant roaming the Arctic in herds! Or to stand in the woods in Michigan and have the skies darkened by flocks of the legendary passenger pigeons? Well, yes, it would be, but surely we're not going to put millions or even billions of dollars into bringing back extinct species just so we can gaze at them?

It's easy to argue that every dollar that goes into bringing back extinct animals is a dollar that could have been spent on conserving the species we already have. Paul Ehrlich, the beleaguered author of *The Population Bomb*, calls it "a fascinating and dumb idea . . . it would divert us from the critical work needed to protect the planet."[25] (Note that at least he acknowledges the appeal of bringing back the mammoth.)

One counterargument is that private donations make up much of the money for the mammoth and the pigeon; there's been little competition for government money. Even so, that money could have been spent on conservation.

But the central arguments for de-extincting (de-extinguishing?) the passenger pigeon and the woolly mammoth are ecological.

The mammoth is the de-extinction project that gets all the attention and fires people's imaginations—mammoths walking the Earth again! Once they're re-created, they could live in Pleistocene Park in northern

Russia.[26] This is an ambitious attempt to take the far northeastern part of the country back in time, populating it with animals most closely resembling the Pleistocene fauna in residence at the end of the Ice Age, ten thousand years ago. The area was then grassland, called the Mammoth Steppe. The founder of Pleistocene Park, Sergei Zimov, would love to add a herd of woolly mammoths. Why? He's convinced the big herbivores engineered the landscape, changing it gradually from forest to grassland by grazing, turning over the soil, and scraping bark off trees as bison do. Or even knocking trees down, à la elephants. The idea is that hooves disturb the surface, allowing cold air to penetrate to the permafrost below, keeping it frozen and holding on to its carbon dioxide. As it soon as it melts, that CO_2 is added immediately to the atmosphere. So mammoths in Pleistocene Park might help reestablish grasslands. And stabilize climate. That was the original goal.

George Church at Harvard and MIT is the central scientist in this de-extinction project. It's a long slog and they've altered their goal slightly: now they're not calling the animal they're trying to create a "mammoth," but instead an "arctic elephant." Or a "mammophant." And the goal is not necessarily to refurbish the tundra, but to extend the range of elephants northward, thus creating new and possibly safer habitats for them.[27]

Church has been at this since 2007, and the list of what remains to be done is about as long as the list of achievements so far. They have an excellent mammoth genome, they've actually inserted some mammoth genes into the genome of the Indian elephant (its closest relative), but that's it so far. Having assembled bones and flesh from carcasses that have just thawed out of the permafrost, they have established that the mammoth and the Asian elephant are 99.6 percent genetically identical. But while that 0.4 percent difference doesn't sound like much, it contains a significant number of as-yet-unidentified mammoth genes thought to be crucial. So there's still much work to be done.

It's interesting that among the investors in the mammoth projects are both a company called Colossal and the CIA (supposedly their interest lies in the technologies that might spin out of this project).[28] Cynics suspect that the company might think that the ideal number of reborn mammoths would be two. They'd make a pretty fantastic zoo exhibit.*

But while the mammoth gets the attention and some money, the passenger pigeon is a close second. Back in the day it would have been a sensational experience to see their unbelievable numbers, estimated to be 3 to 5 billion birds, occupying the forests of northeastern North America. They were social birds that nested in spring and then abandoned their babies two weeks after birth to join other adults in long flights searching for food. These flocks were so big that when they descended into the forest they would break off branches with their weight, destroy foliage, and cover the forest floor with their guano. A mass landing like that would change that part of the forest completely, opening up sunny areas on the ground and enabling a new generation of plants.

John James Audubon, the author and illustrator of the legendary *Birds of America*, described one flock along the Ohio River this way: "The air was literally filled with Pigeons; the light of noon-day was obscured as by an eclipse." Audubon tried to keep track with a pencil and paper but soon gave up—there were just too many. He estimated he'd seen a billion birds that day.[29]

Ben Novak is lead scientist at Revive & Restore, a de-extinction organization that promotes "the incorporation of biotechnologies into standard conservation practice."[30] He calls the passenger pigeon "the dance partner of the forest" because of the impact it had, constantly forcing the forest to rejuvenate.[31]

The technology is the issue here. There are many passenger pigeon skins in museums, so getting bits of tissue for DNA analysis hasn't been difficult. Then, like the mammoth, genes unique to the passenger pigeon are being identified. It's necessary, again like the mammoth, to find the most closely related living species to hatch the eggs. In this case it's the

---

* Fantastic also is the idea of bringing back a Neanderthal. Church made reference to that idea in an interview a few years ago, but he was not advocating we do it today with the Neanderthal genomes on hand, but rather stating that the technology could make it possible.

band-tailed pigeon, a bird that inhabits the American Southwest. But engineering these eggs to contain passenger pigeon genes is a huge challenge. The near-term goal is to gradually, over a few generations, push closer and closer to a bird that could reasonably be called a passenger pigeon. But if you're going to try to rejuvenate eastern North American forests, that's just the beginning—you'd need thousands of birds.

Novak isn't deterred by the numbers. Because the parents leave the nest before the offspring can even fly, he argues that most of the young's social smarts derive from being with other offspring once they've all fled the nest and started to socialize. Novak imagines establishing two giant aviaries in the United States, one close to their original nesting area and one far away where they used to land and feed. Then train homing pigeons, appropriately colored to look like passenger pigeons, to fly from one aviary to the other accompanied by the inexperienced pigeon young. Scientists not directly involved in the project see unacceptable cruelty in birthing a passenger pigeon that is then unable to form adequate social bonds. It's not even clear that today's forests of the northeastern US provide suitable habitat, and tomorrow's, as they experience climate change, might make them even less hospitable to passenger pigeons.

Looking at the challenges facing the technology, the enormous time frame, and the unanticipated problems, I think this is one technology enabled by genetics that isn't going to advance the cause of conservation. Nonetheless, even a couple of passenger pigeons would at least create a sensation—for a while.

Authenticity Counts

Nature is an all-senses experience. If you need to be immersed in all of nature to derive the biggest benefit, how do you deliver that experience to people living on the fifty-fourth floor downtown, kilometers away from any place you could argue was natural?

Could technology provide the answer to this question? Maybe the "metaverse," another place where we might live in 2050, could make this happen. Tech writers everywhere are now laughing that I would take the metaverse seriously even for a second. Acknowledged. So far, it's clumsy, oversold, and a money drain. But let me fantasize for a moment and try out the idea of bringing nature to people instead of people to nature—virtually. Besides, if it's not the metaverse, maybe it could be something not unlike it. Facebook's (actually Meta's) Mark Zuckerberg is generally given credit for bringing attention to the idea of the metaverse when he pledged that Facebook was going to throw itself into helping create "an embodied Internet that you're inside of rather than just looking at."[32]

If you've had any experience with the Oculus helmet or another virtual reality (VR) setup, you'll know that the experience is compelling but not convincing.* But given the amounts of money invested in the idea of the metaverse, the technology is quickly getting better. Connectivity is faster. Sound and image are finer grained than ever. Assemble all the technologies that could work together to make a virtual world that's real, natural, totally believable, and totally convincing; have them create a "digital twin" of the place you live; then you can go into the virtual twin, as an avatar, hang out for a coffee, do some business at the bank, then go for yoga. Not here—in the "metaverse."

Years ago, a friend of mine joined Second Life, the AA-baseball version of a major-league version of the metaverse, though pretty remarkable in its day. It was a virtual world where people presented as avatars, and did the sort of things they already did in the real world: converse, party, buy and sell, even invest. Businesses set up offices in Second Life and made money.

I remember hanging out on my friend's Second Life back deck and spotting a neighbor at her door a couple of backyards away. We called her over to talk to her, and as I remember, she crossed over to the yard immediately behind us to talk. I don't remember what was said (were we

---

* Although I admit that I was once touring a virtual office space, and when invited to step off the balcony and fall to the (virtual) ground, I could not do it! All the evolved cautionary mechanisms in my brain overrode the fact that I *knew* I was on a level surface.

talking about opera?), but I do remember being kind of amazed that this had just happened. I had no idea who this person even was.

Now imagine, rather than looking at a screen image of Second Life, actually entering it and walking its streets. You can see 360 degrees. You can meet old friends, shoot baskets at the corner. It's very close to the experience of the life you lead here, in the real twin. As it gets more refined, it passes even closer inspection. You can even feel the little bristles on the zucchinis as you pick them with your haptic glove. It's so convincing.[33]

It might seem from the research I just cited that creating a virtual nature that's indistinguishable from the real thing would be a huge challenge: Can the atmosphere of the forest, the airborne molecules, be duplicated technologically? Could you get the same improvements in health and well-being? It would be complicated: you'd already know that it *should* be beneficial, thus creating your own positive placebo effect, but you'd also know at every moment that you aren't actually in "real" nature. Maybe the two would cancel each other out.

But there's experimental evidence that virtual environments have many, if not all, of the beneficial effects of actually being there. However, as they say, the devil is in the details. For instance, in one setting the effect that virtual nature had on participants was dependent on which natural scene they liked the best.

Another recent study revealed that having people walk down a virtual, sandy, palm-tree-lined path to a virtual ocean shore improved their emotional states and lowered stress, with one catch: stress was lowered when the experimenter controlled the walk, but not when the participants controlled it themselves.[34]

There's still much to learn then, and in the end there might be individually chosen virtual reality scripts for stress, mood alteration, lowering blood pressure—any number of measures. Once the metaverse is established, it could indeed become a place where you might consider you're "living." Maybe you could become a metaverse "snowbird" and take a couple of months in the winter, when it's hostile outside, to hang out with friends in one of the many metaverses you've established your avatar in. You do yoga at an Indian ashram, you throw pots in San Francisco, you sit in front of a waterfall. All without leaving your home. It can be risky—apparently a guy stabbed himself on a glass table while dancing around the room. Zuckerberg has said the true birth of the metaverse will be when "immersive digital worlds become the primary way we live our lives and spend our time," but the size of the Zuckerberg audience is shrinking.

As far-fetched as this might seem, if significant numbers of people explored the metaverse, as are now using cell phones, it would have a huge impact on life in the city of 2050.

Exurbia Max

There's an array of possibilities, from smart city to nature city to virtual city, and of course myriad other possibilities dictated by local features and cultures. But there is one more, even further-out-there possibility. Could we live in space?

Science fiction has told us repeatedly it will be routine, and even glamorous. Doomsayers point out that eventually something—an asteroid strike, a nuclear holocaust—will eliminate us, and unless we've taken the dramatic step of moving off the Earth, then we're done. But what would life in space or on another planet really look like? One thing is for sure: the farther from Earth, the more daunting.

Life in low-Earth orbit, like the International Space Station (ISS) at 400 kilometers or close to 250 miles above the Earth, is not as glamorous as it might seem and actually is, with time, debilitating. A life in zero gravity being bombarded with space radiation has serious health impacts. The ISS is also a constrained environment in a hostile place, conditions that can trigger conflict and affect mental wellness.

Scott and Mark Kelly, identical twin American astronauts, provided an opportunity for a unique controlled experiment when Scott spent nearly a year at the ISS while Mark remained earthbound. After Scott returned to Earth, the data from his self-administered medical tests while in orbit were compared to his brother's.[35] There were significant differences, including changes in Scott's DNA (thousands of them) and his microbiome (the suite of microbes in his gut). The genetic changes affected some genes that repair DNA, an alarming twist given that he had absorbed a dose of radiation forty-eight times the average on Earth. Some of these changes were reversed once Scott had been back on Earth for a while, but some persisted. And he was in space for only 340 days, not several years.

Gerard O'Neill was never a celebrity like space billionaires Elon Musk or Jeff Bezos, nor did he command companies that built spaceware, but for a time back in the late 1970s the late Professor O'Neill captured the imagination of space buffs by proposing that people would one day live in space colonies far from Earth. A physicist at Princeton University and author of the book *The High Frontier*, O'Neill argued that constructing space colonies would be easier, cheaper, and create better living environments than trying to build a civilization on Mars. His idea was that colonies would be positioned at or near Lagrange points in space—places where some combination of the gravitational forces of the Earth, Moon, and Sun enable spacecraft to arrive, stop, and remain roughly in the same place essentially forever. As O'Neill put it, "the ultimate size limit for the human race on the newly available frontier is at least 20,000 times its present value."[36]

O'Neill colonies—as they were called—seemed very attractive places to live. They were gigantic cylinders, sometimes a few kilometers long, rotating slowly to generate the equivalent of Earth gravity. Given the enormous energy cost of launching building materials from Earth to space, O'Neill colonies would be constructed from materials mined and processed off-Earth, probably on the Moon or an asteroid. With abundant solar energy, agriculture would thrive, growing food for a population of maybe a million. It all looked so peaceful, almost bucolic, and you could enjoy un-Earthly adventures, like being able to fly near the axis of the cylinder where the gravitational force was lower.

O'Neill was a member of that camp urging us all to think seriously about getting off the face of the Earth. Some are still making this argument. One is Amazon's Jeff Bezos, who, while he isn't competing in billionaire-promoted suborbital flights, is dreaming of space colonies.

He envisions people being born in the colonies and only visiting Earth on vacation. But more intriguing are the plans of an organization called the National Space Society in the US, which endorses a rewritten version of O'Neill's original plans that brings colonies much closer to Earth, and therefore creates a much more rapid deployment.[37] Briefly, they plan to orbit roughly 500 kilometers or 310 miles up, about 100 kilometers or 60 miles higher than the ISS, over the equator. In that locale the Earth's magnetic field shields a colony from almost all space radiation. So much less protective shielding from space radiation would have to be hauled from the surface of the Earth into orbit, a very expensive and time-consuming task. The thinking is to start small with a colony about the size of a cruise ship. At that size a colony would have to rotate about four times every minute to create Earthlike gravity by centrifugal (or more accurately, centripetal) force, a disorienting sensation to say the least but one that proponents claim humans could get used to. This proposal's outstanding advantage is the colony's proximity to Earth.*

For instance, the Moon is 720 times farther away. Living on the Moon will be necessary if there's industrial activity established there, like mining, which is one of NASA's goals. My largely science fiction–driven view of life on the Moon is not a pretty one, but I'd argue it's grounded in reality. First, the daily rhythm of night and day doesn't exist. Well, it does, but on a dramatically longer time scale. Days and nights are each fourteen days. At least in the land of the midnight sun here on Earth, there's seasonal relief. Not on the Moon. There are deadly cosmic rays and ultraviolet light from the Sun, unfiltered by the atmosphere, making living underground almost mandatory. Perhaps cool residences can be established in one of the long lava tubes spotted by the Lunar Reconnaissance Orbiter that run underground from the Moon's surface. Such housing would likely be windowless.

---

* Some aren't intimidated by great distances from Earth. Finnish scientist Pekka Janhunen has made the case for establishing a set of O'Neill colonies orbiting the asteroid Ceres, in the asteroid belt between Mars and Jupiter. The asteroid would serve as a source of raw materials for the colonies.

Gravity on the Moon isn't as negligible as it is in low-Earth orbit. It's one-sixth that of Earth's, which is what allowed the astronauts half a century ago to hop around the lunar surface. But spending too much time in one-sixth gravity, as you would if you lived on the surface of the Moon, might make it impossible to return to Earth. If you did, you would need medical support and long-term care for your lunar-accommodated body. And while NASA is overtly concerned about the mental strain of being part of a small crew sitting essentially passively for months at a time on the way to Mars, living on the Moon, even with many people around, would be a similar situation.

One sci-fi scenario that caught my attention was the claim that living and working on the Moon would demand extraordinary measures to ensure safety in a hostile environment. The kinds of people comfortable with enforcing rules might thrive in such an environment and that might not be the best recipe for social peace. But who knows? No one's living on the Moon yet, but it wouldn't be surprising if they were ten years from now. They would always be remembered as the first to live in space. If they were to be the last, that would be a sorry end to generations of science fiction fans. Anyway, the latest estimates for sending people to the Moon center on the next decade.

What about humans on Mars, Elon Musk's (and others') dream? Mars is about 55 million kilometers (33.9 million miles) farther away than low-Earth orbit. In any case the issue is not just the destination, but the commute. Every risk to human health and well-being that has manifested on the International Space Station would exist in spades on a trip to Mars. According to NASA, that includes "altered gravity, radiation, hostile environment, isolation/confinement, and distance from Earth."

There would be three versions of gravity: excess G's upon launch, zero G's for the bulk of the trip, and Martian G's, a little more than a third of the Earth's gravity. Recovering from the zero G of the trip might be pretty tough—a Mars voyage is expected to last seven months. But living in one-third gravity on the planet's surface will not be healthy, either, although the exact toll it will take is unknown. Radiation risks are equivalent to those on the Moon, and currently there is no available shelter other than within the walls of the spaceship. Space is hostile. And the last two—isolation/confinement and distance from the Earth—are a one-two punch to human equanimity and mental health. Remembering that Mars and the Earth are constantly moving, a round-trip to Mars could take about two years: six to eight months to get there, nine more months to explore and hit pause to wait for the two planets to move into the right positions relative to each other, and the concluding six-to-eight-month return trip. Now, if you read or saw the movie version of *The Martian*, just transform that story from "surviving long enough to get back to Earth" to "living there permanently." It doesn't sound that attractive to me. But—and this is the sentiment of many who'd like to go there—Earth in 2050 and beyond might not be, either. If going to the Moon ends up being in the 2030s, current estimates are that Mars could be ten to fifteen years after that.

Elon Musk, for one, is not deterred from getting to the red planet, even with the difficulty of landing a huge spaceship on Mars. Given the near uselessness of parachutes for that task, Musk has said, "I want to die on Mars. Just not on impact."[38]

CHAPTER 11

Tech for the Planet

The late Stephen Hawking insisted that we must move to space before Earth becomes unlivable. Jeff Bezos has agreed, saying, "We are in the process of destroying this planet."[1] The Mars Society is eager to establish "a new branch of human civilization" on the red planet. While you could imagine the trigger for these statements being a nuclear holocaust, or a pandemic much more severe than COVID-19, all that's really needed is uncontrolled climate change. Rising oceans, scorching heat waves, the deaths of insect pollinators, and the failure of agriculture would all occur in some of the worst climate scenarios. Could there be a technological solution?

In 2020, global carbon emissions dropped by 6.4 percent because of the global pandemic. That's about double Japan's annual emissions. But it didn't last long—in 2021, carbon emissions were at the highest levels ever. That isn't surprising, because even though some metrics of climate change appear to be encouraging (the International Energy Agency predicts that global carbon emissions will peak by 2025, bolstered by an incredibly rapid uptake of renewable energy), the picture looks bleak unless efforts to reduce carbon dioxide emissions are redoubled, then redoubled again. Otherwise, the future Earth is going to look very different (and not better!) from the one we're used to. Sustainability is the way to go, but if there's too much foot-dragging, could we turn to technologies to deploy in addition to cutting back our emissions?

The Industrial Revolution? Amazing! The Green Revolution? Saved billions of human lives! However, both have contributed to the record levels of carbon dioxide in the atmosphere—more than we've had for the last two million years and it's hotter now than any time in the last 125,000 years. And it's going to get worse before it gets better, "getting better" meaning when carbon dioxide levels finally start to decline because we have stopped flooding the atmosphere with it.

We are familiar with the myriad changes we would need to lessen the risk of catastrophic climate change, everything from personal choices, like switching your car from internal combustion to electricity or even to an e-bike, simply commuting less, reinsulating homes, and eating less meat or any emissions-intensive food. At the corporate level, there are plentiful promises to become net-zero by 2040 or 2050. But there are serious doubts that these and other measures, even together, will allow us to remain below a 1.5-degree Celsius (or even 2-degrees Celsius) increase in global temperature since the late 1850s. That's when the impact of the Industrial Revolution first began to be felt.

According to the Intergovernmental Panel on Climate Change (IPCC), we have less than a decade to try to rein in temperature increases to less than 1.5 degrees Celsius, which would prevent the exposure of hundreds of millions of people to *extreme* heat waves and water shortages; ensure the survival of some coral reefs; perhaps avoid the melting of the West Antarctic Ice Sheet; and preserve habitats for as many as 50 percent of all vertebrates. Crashing through the 1.5-degree Celsius barrier is likely; some forecasts suggest even 2.0 degrees Celsius will be reached by 2050.

The rapid deployment of renewables like solar and wind (at a speed that few predicted) is already having a positive effect, but it's not enough.

Climate change is a life-or-death situation for millions, and as is sometimes said, physically possible but politically extremely difficult if not impossible. So if the combined efforts of individuals, nongovernmental organizations (NGOs), and corporations fall short, what else can we do? There are two technologies in particular that are aimed directly at climate change. Both are controversial. One involves removing the greenhouse gas carbon dioxide from the atmosphere, the other shielding the Earth from the Sun.

A Window of Opportunity

There are several approaches for removing carbon from the atmosphere. The simplest, of course, is planting trees. Trees have hundreds of millions of years' experience in downloading carbon dioxide to fuel photosynthesis (sunlight + water + CO_2 = oxygen + sugar). But there are significant risks with relying on this form of carbon removal: there is no guarantee that the seedlings will mature into carbon-gobbling full-size trees. Drought, insects, a cold snap, or flooding can kill any newly planted slip of a tree. Even if those trees do mature, they could be cut down or burned, meaning that the stored carbon would quickly be released back into the atmosphere. The number of trees that would have to be planted to effectively absorb the carbon dioxide would have to be in the trillions. After all, on average a tree absorbs about 22 kilos (48 pounds) of CO_2 a year, equivalent to the emissions produced in the farm-to-table production of three average hamburger patties![2] So the benefit to the climate of that big maple in the front yard was negated by that modest neighborhood barbecue you hosted. Also, climate change itself will throw curves by increasing the frequency of tree-killing wildfires, rendering formerly ideal tree habitats uninhabitable as temperatures rise or necessitating a completely different set of species to be planted.

A related technology that relies on photosynthesis is called bioenergy with carbon capture and storage, or BECCS. It's straightforward enough: grow inedible crops or grasses, burn them to generate energy, and at the same time capture the carbon that's emitted and bury it underground. In the decades to come, especially in the second half of this century, it's assumed or hoped that BECCS will play a significant role in carbon removal. However, it's not without controversy. One issue is uncertainty over its value: each step in the process is fairly well understood, but whether the energy generated will compensate for the energy expended in growing, collecting, packaging, and transporting the crop to be burned is uncertain. Taking over potentially agriculturally productive land or natural habitat to grow the biomass would negate much of the value.

There is also widespread concern that any dependence on a technology that hasn't yet proven itself could reduce concern over limiting carbon dioxide emissions. Why bother cutting back on fossil fuels when we can just remove the CO_2? This issue, more psychology than technology, is commonly called a "moral hazard," a simple example being the store owner who takes out theft insurance and then is less careful about locking up her shop at night. It's not surprising it's a default position considering the efforts fossil fuel companies (and politicians) have made in the past to obstruct moves to curb carbon dioxide emissions.

A second technology was pitched by American environmental engineer Klaus Lackner, founding director of the Center for Negative Carbon Emissions at Arizona State University.[3] He developed what's called the "Lackner tree," which stands about 10 meters or 32.8 feet in height, 1.5 meters or 4.9 feet in width, and contains stacks of synthetic leaves made from tile coated with a CO_2-absorbing resin. The resin works on a wet/dry cycle, absorbing CO_2 when dry and releasing it when wet. Once the tiles are saturated with CO_2, they are moved to storage, where eventually they're humidified to release the gas. The discs are claimed to be a thousand times more efficient than real leaves (although they'd have to be located in low-humidity environments). Sounds puzzling, but a natural tree's business is to use carbon dioxide to grow and doing that requires a complex metabolism that to some extent limits carbon dioxide uptake.

The Slow Approach

Two other technologies stand out for the attention they have attracted, both positive and negative. One is called air capture, or direct air capture. The other is geoengineering.

Ten years ago, it was hard to find a scientist who would take air capture seriously; now there are many. This approach can be accomplished in different ways, but the idea is to trap carbon dioxide, either as it emerges from the industrial pipe or by grabbing it directly from the ambient air. Both are tricky technologically and economically: not only does the capture of carbon dioxide (sparing other gases) have to be highly efficient, but so does separating out the CO_2, finding a way either to use or store it, and then recycling the material used to capture it.

Capturing carbon dioxide just as it's leaving the industrial stack is effective in one sense—it captures about 90 percent of the gas—but fewer than forty are fully operational, and maybe another thirty or so are at some stage of development.

The Center for Climate and Energy Solutions in Arlington, Virginia, estimates that this technology has the potential to remove about 14 percent of the CO_2 that we need to eliminate to achieve net-zero emissions by 2050.[4] Not enough.

Air capture has faced years of skepticism (if not outright ridicule), but now has been acknowledged by the IPCC as an "essential" part of reaching climate goals. Nonetheless, there is still caution. While the technology can probably work, it might never be cost-effective or scalable fast enough, and again, there's the "moral hazard": it threatens to weaken the will to reduce emissions.

A more ambitious variation of capturing CO_2 at the stack is to pull it out of the atmosphere long after it's escaped. This technology, called "direct air capture," is a bigger challenge. While the concentration of CO_2 coming out of a stack might be 5 to 20 percent, once it's dispersed in the atmosphere it is only 0.04 percent, a very elusive quarry. An ironic twist is that reports of progress in direct air capture can be, at the same time, admissions of the near impossibility of the task. For instance, in

2017 Climeworks AG in Switzerland announced they had become the first company to capture CO_2 on an industrial scale from the air and sell it.[5] Their goal for this pilot plant was to capture 900 tons of CO_2 annually, about the amount vented from two hundred cars. To reach their ultimate goal of removing 1 percent of the global atmospheric CO_2, they would need to build 200,000 similar plants. That's success? It was—in the sense that it signaled possibility.

Climeworks didn't stop there. They've now built a much larger plant in Iceland, capitalizing on the abundant sources of geothermal energy to multiply the capacity of their original plant by four without having to use fossil fuels. This plant, called Orca, is one of fifteen of varying sizes built by different companies worldwide.

Some of those companies have similar ambitions around growth. Carbon Engineering, a company that began with a trailer-sized setup in Calgary, Alberta, more than ten years ago, has now embarked on building a commercial plant in Texas that aims eventually to capture a million tons of CO_2 a year.[6] Their technology, straightforward but costly at industrial scales, begins with atmospheric air passing over surfaces with potassium hydroxide flowing over them. The solution captures CO_2 and processes it through a series of concentrating and compressing steps so that pure CO_2 gas is the end product. The chemicals crucial to the process are then recycled to capture more.

It can be done anywhere. Because carbon dioxide, no matter where it's emitted, becomes part of the general circulation of the atmosphere, there are no geographic constraints on the location of a plant: if solar powered, then wherever there is the most sunlight; if wind powered, the same; or, as Orca shows, if geothermal, then put it 50 kilometers or 31 miles from Reykjavik. If it were convenient to assemble the construction materials in the Sahara Desert, they could set up there.

If either Climeworks or Carbon Engineering achieves the volumes of captured CO_2 they're aiming for, the issue then looms large of what to do with millions of tons of carbon dioxide. An irony for a technology hyped for removing a gas produced by fossil fuels is that one of the first reservoirs for carbon dioxide recovered from a coal-fired power plant in

Saskatchewan was to ship it by pipeline to oil fields to inject back into the ground to enhance the oil recovery. That is no longer the focus. Most contemporary projects aim either to sell the CO_2 or dispose of it, usually by burying it. Deep underground.[7]

Although Climeworks' small plant in Switzerland sells their relatively meager amounts of CO_2 to a greenhouse to enhance vegetable production, in Iceland they've opted for the deep underground route. Orca pumps the collected CO_2 deep into the ground, where it is deposited as granules of calcite in basaltic rock.

Carbon Engineering is using a similar approach. Working with Occidental Petroleum, they plan to build their megaton plant in Texas in what's called the Permian Basin, the huge sedimentary formation underlying the western half of Texas and southeastern New Mexico. The area is full of oil and natural gas. (It's not just the Permian Basin—the US has the underground space to store more than 8 trillion megatons of carbon. If you like playing the "yeah, but how many car exhausts does that neutralize?" game, it's the emissions from 35 billion vehicles over fifty years.)

Occidental has years of experience in carbon capture and employing carbon dioxide to boost the recovery of oil. But Occidental is putting money into this new direct capture plant, one that will ultimately be bigger than Orca and will possibly be twinned with another of similar capacity, and one that will inject the collected CO_2 deep into porous rock trapped forever (hopefully) under an impermeable cap.

A third company, 1PointFive, is building a system to generate low-carbon jet or diesel fuel.[8] Making jet fuel is much more complex than piping CO_2 into the ground. It begins with the CO_2 collected from the atmosphere, uses renewable electricity to split the hydrogen from H_2O, then combines the hydrogen with carbon dioxide to create hydrocarbons. Ideally a process like this could provide the "drop-in" fuels that airlines are so avidly waiting for.*

Why is making jet fuel this way important for climate change?

---

* The European Union's SUN-to-LIQUID project has just produced the first carbon-neutral jet fuel and plans to start industrial production in 2023.

Standard jet fuel is derived from oil that has been buried in the ground for at least a hundred million years. Burning it releases carbon dioxide that is, in effect, "new"—an addition to the modern atmosphere. But recycling carbon dioxide from the production of oil and gas is different. The CO_2 does not escape and so does not add to the atmospheric burden of the gas. Instead, it's processed so that it can be used again. There's no reason, if there was adequate technology, that you couldn't keep reusing carbon dioxide as jet fuel over and over. The chemistry works, but whether the process is economic and can be scaled up fast enough is another question.

Geoff Holmes, business development manager at Carbon Engineering, has seen direct air capture graduate from inattention to excitement. But he is still cautious: "We will have succeeded when a direct air capture facility will be as unremarkable as a water treatment plant."[9]

It's not yet clear that any of these technologies will reach a level of economic viability such that they'll be embraced on a global scale. That would be quite an embrace, too: by 2050, it's been estimated we'll need a hundred or even a hundred and forty times more direct air capture plants than are even in development.

In 2016, climate scientists Kevin Anderson and Glenn Peters argued in *Science* magazine that future climate scenarios place unjustifiable weight on technologies like direct air capture and other ways of drawing carbon dioxide from the atmosphere collectively known as "negative-emission technologies."[10] The fact that these largely unproven technologies are critically important to hit climate targets could of course cut either way. If they work, great! We will hit the targets and won't have had to pour money into other efforts, like retrofitting housing and eliminating fossil fuel use. But if they don't work, we've counted almost exclusively on technologies that have failed and the climate situation would be dire. In their words, "Negative-emission technologies are not an insurance policy, but rather an unjust and high-stakes gamble."

Counter that argument with this fact: most of the burden of extreme climate change is borne by middle- to low-income countries, sometimes called the Global South. They don't produce most of the CO_2 or methane; the Global North does. It is true that a single round-trip transatlantic flight releases as much CO_2 per passenger as a year of commuting 35 kilometers or 22 miles daily in a Prius, but it's also more CO_2 than the *annual* emissions of an individual living in sub-Saharan Africa. Ramping up the development and construction of more air capture *around the world* might be a small step toward reparations.

Faster—but Better?

In April 1991, Mount Pinatubo in the Philippines, long known to be an active volcano, suddenly erupted in a violent eruption of steam, ash, and lava. It was just the beginning. In June, after months of building pressure, an eruption shot 5 cubic kilometers or 3.1 cubic miles of gas and ash up 35 kilometers or 22 miles into the atmosphere. The mountain collapsed, and over the next few weeks heavy rains, ashfall, and flows of volcanic debris killed about eight hundred people. A tragedy on the ground, although safety precautions put in place before the eruption limited the number of lives lost.

But the eruption had an intriguing impact on the atmosphere. The sulfur released by the volcano circulated around the world in the upper atmosphere and caused global temperatures to drop by 0.5 degrees Celsius. The sulfur released from the volcano combined with water to form fine droplets of sulfuric acid; the droplets partially intercepted or reflected incoming sunlight. This layer of droplets had multiple effects on the Earth's climate, but in particular that reduction of temperature in the lower atmosphere was a brilliant natural demonstration of the power of sulfur-containing aerosols to reduce incoming solar radiation and trigger a drop in temperature. Now, switch a volcano to a high-flying airplane releasing similar sulfates into the atmosphere and you have a potential mechanism for turning down the Earth's temperature.[11]

This is only one proposed geoengineering scheme. Another is space mirrors—giant reflectors launched to a location called Lagrange 1, about a million miles from Earth.[12] Balanced by the Earth's and Moon's gravitational forces, they would essentially block incoming sunlight or reflect it back into space.*

Whitening clouds over the oceans to make them brighter, and hence more reflective, is another geoengineering technology. This would involve spraying particles into clouds, ramping up the number of cloud droplets as a result, and the newly brightened clouds would then turn back some of the incoming solar radiation. Something like a doubling of droplets in clouds could exert a significant dampening of solar radiation, but there are many unanswered questions about cloud droplets and how they form. Even whitening the roofs of buildings has been suggested. To me, all of these have a slight air of desperation about them (you know you're speculating when some of the most prominent ideas for increasing cloud droplets over the oceans utilizes robotic seagoing vessels).

Each of these, regardless of how feasible they are, might theoretically be useful, but the high-altitude version, injecting a blocking layer of sulfates into the atmosphere, has attracted the most attention, both positive and negative.** Its status is uncertain: while describing direct air capture as "essential," the IPCC virtually ignored geoengineering in its 2018 report on efforts to limit temperature rise to 1.5 degrees Celsius.

Is it even doable? There have not yet been enough experiments to answer that question. But there are enough doubts and questions to have made this potential technique even more controversial than direct air capture was a few years ago. From this point on, I'll use the term *geoengineering* to refer only to this atmospheric blocking of solar radiation.

Once again there is, of course, the so-called moral hazard, the risk that a demonstrated means of reducing global temperatures would encourage corporations, political parties, and countries to scale back their efforts to curb emissions. In fact, because geoengineering is effective much sooner than direct air capture, the moral hazard seems more urgent. But a second

---

* There is already a solar observatory parked at L1.

** As an example, this phrase was the title of one section of a critical article: "Hothouse Earth or Shithouse Earth?"

worry unique to geoengineering is something called the "termination problem," and it is a clear mark of differentiation from direct air capture.

If direct air capture were extensively deployed then suddenly stopped, CO_2 accumulation would start to rebound and temperatures would rise. But the rise would be gradual and not immediately catastrophic.

On the other hand, if geoengineering were suddenly stopped, atmospheric sulfates that had already been injected in the stratosphere would disperse quickly, giving solar radiation unimpeded access to the lower atmosphere. If efforts to reduce emissions had been relaxed and CO_2 levels were high, climate hell would break loose. If geoengineering had been under way for, say, two decades, and then were to stop, temperatures would rise to where they would have been anyway, but would get there much faster. Calculations have suggested that rate of temperature increase might be ten times what we're experiencing now. Rates like that would severely challenge the ability of all life on Earth to adapt quickly enough.

This is exactly why efforts yielding significant results in both carbon dioxide emissions and capture would have to be in place *before* any such intervention takes place.

On the surface the potential looks good. Something like a 2 percent reduction of incoming solar radiation could neutralize a *doubling* of atmospheric carbon dioxide. The effects would not be felt equally across the globe. The tropics would be cooled much more substantially than the poles (which are already experiencing intense warming). Also, the atmosphere would become more stable and precipitation would fall, especially in the tropics.

The combination of a cooler, drier tropics is not something we would want to see. This is another argument for reducing emissions aggressively before contemplating geoengineering. Even if we blocked as much solar radiation as feasible, we would not be able to restore either precipitation or temperature to pre-industrial levels equally around the world. Efforts to rebalance temperature would lead to precipitation declines but restoring precipitation could worsen warming, at least in some parts of the world.

Of all the future technologies I've discussed in this book,

geoengineering stands out for the abundance of social and political issues it raises. But there are also technological issues that complicate the idea of deploying aerosol particles into the stratosphere.

First, which sulfur-containing (or other) compounds are best, where "best" includes a number of factors: staying power in the upper atmosphere, ability to reduce incoming solar radiation, and ease of transport. Speaking of transport, high-flying aircraft or tethered high-altitude balloons are judged to be the best candidates for courier, but the aircraft would have to be custom-built to reach the necessary altitudes. The ideal staying power and effectiveness depend on the size of the particles, and their size is difficult to ensure once they're released. One advantage/disadvantage is that particles would generally last no longer than two months in the stratosphere, a good thing if something goes wrong but bad because that short life necessitates never-ending deliveries of chemicals to the upper atmosphere.

Geoengineering also risks something these other technologies don't: geoengineering wars. Country A opts for much cooler temperatures and so launches a huge geoengineering effort. Neighboring country B is not happy—they want higher temperatures for their crops—so they look for ways to stop country A. If diplomacy fails, then who knows what would ensue? It could be as precarious politically as climatically.

The Experts Disagree

Ensuring that geoengineering would be universal and equitable is an issue. A short paper in the journal *Science* argued that no experimentation on geoengineering should go ahead until there's better public information distributed more widely. A research team led by Jennie Stephens of Northeastern University took no prisoners in their approach: "Advocacy for solar geoengineering research continues to be dominated by white male scientists from the Global North funded by tech-billionaires and elite philanthropy."[13] Not surprisingly, they concluded by advocating for international oversight of geoengineering, making it a UN mandate, independent of national policy. However, there is also research suggesting that the countries of the Global South are open to geoengineering. These authors at Northeastern, as strident as they were, didn't go so far as to advocate banning geoengineering forever, even though there are those who do.

Extreme opinions like "ban geoengineering" were countered in an editorial in the same journal by Edward Parson, an environmental law professor at UCLA: "To reject an activity based on harms that might follow is to apply extreme precaution. This can be warranted when there is risk of serious, unmitigable harm and the *alternative is known to be acceptable.* That is not the case here. Rejecting SG [solar geoengineering] research means taking the alternative trajectory of uncertain but potentially severe climate impacts" (emphasis mine).[14]

Finally, Holly Jean Buck at the University of Buffalo pointed out in an article in the journal *Frontiers in Climate* that most of the discussion of geoengineering has focused on the risk of conflict rather than the potential for promoting peace.[15] While it's easy to be cynical about prospects of peace at the best of times, she argues that ignoring any possibility that geoengineering could build, rather than destroy, relationships should be taken into account when deciding its future.

Calling it "weather modification" immediately brings back memories of the United States seeding clouds to promote rain and disrupt North Vietnamese troop movements during the Vietnam War. Instead, Buck argues that it's possible that the Northeastern University group's concerns about Global North domination could be mitigated by the "peace not war" possibility. She imagines a scenario in which taking steps to lower the global temperature by 0.5 degrees Celsius (especially if we have been late to act to cut emissions) could have humanitarian benefits not limited to a single country. But in the end, "who" is doing the geoengineering is just as important as "how."

David Keith of Harvard University has been involved in the geoengineering debates from the beginning (he was also the founder of the company Carbon Engineering). Keith has many sensible things to say about geoengineering, specifically the kind I've been talking about: injecting sulfates into the upper atmosphere. In an article published in the *New*

York Times titled "What's the Least Bad Way to Cool the Planet?" he points out that cutting emissions is crucial but insufficient, because the CO_2 already deposited in the atmosphere will continue to exert its greenhouse effect and polar ice caps will continue to melt—even if we meet the extremely difficult goal of no net emissions by 2050.[16] More has to be done.

But how? Keith admits that direct air capture (like Orca in Iceland and Carbon Engineering in Texas) would "trounce geoengineering in a straw poll," while solar geoengineering is seen as a "desperate gamble." Despite his involvement in, and enthusiasm for, direct air capture, he points out that air capture is expensive. Regardless of how effective air capture might become, it will be expensive and time-consuming to set up and run the supply chains necessary for the rate of installation of air capture facilities needed—supply chains that might be as large and complex as those in the mining industry. Geoengineering would be cheaper and faster. Cheap and fast is good, but as a Band-Aid solution only, not a cure.

Keith agrees that the issues of governance raised earlier are important: as he puts it, "Which country or countries get to decide to inject aerosols into the atmosphere, on what scale, and for how long?" At the moment, not enough experimentation has been conducted, or even approved, to shed more light on this question.

Hasten and Caution

You can't help but get the feeling that we should be making decisions much faster than we are, and that the decisions we have made, like pledges to reduce emissions by, say, 2050, are ignoring the possibility that while all this is better than business as usual, it's still not adequate.

As a final note, obviously some people have decided they *will* make decisions faster. At the end of December 2022, a start-up called Make Sunsets revealed that they had launched a pair of balloons carrying small amounts of sulfur particles into the atmosphere from a site in Baja California, Mexico.[17] Luke Iseman, the CEO of Make Sunsets, admitted that it was part science project, part climate activism. There was no equipment on board the balloons to monitor what happened to them or the particles they were carrying; nor did Iseman seek approval from anyone in Mexico or anywhere else. He did succeed in attracting criticism and even ridicule for confirming critics' beliefs that this technology will inevitably attract rogue operators. Apparently, it did.

CHAPTER 12

Meet the Neighbors

This chapter is definitely about technology, but is it about our future? You could phrase the question somewhat differently: It *is* about the future and our place in the universe, but does the subject actually exist? If it does, good or bad, it will have the most enormous impact that our species has ever experienced. Unpredictably enormous. It would seem that if we're going to survive as a species for thousands of years more, perhaps living elsewhere in the solar system, we'd like to know if we're alone. Of course, the topic is aliens: intelligent life elsewhere in the universe.

When it comes to aliens, no matter what age you are, you will have a signature image in mind triggered by the word, perhaps something more insectlike than human, a green-skinned, large-eyed, dome-headed biped, or an inscrutable being inside an armored exterior. The entertainment industry has a powerful hold on the notions of an alien. But there's also science that comes into play that encourages bigger questions than just guessing what they'll look like. Questions like: How likely is it that they're out there? How do we look for them? And if we find them, what do we say—or do we dare say anything?

There's a natural step forward from technologies that I've written about that are already proving themselves, to those that hold some future promise, now to technologies applied to a subject for which there is no evidence at all! It's strange, isn't it? If there are no other intelligent beings, then what I'm about to say is completely forgettable. If there *are* others, human life will be irrevocably changed. Much thought has been put into how best to communicate with (hypothetical) aliens, either by detecting their signals or sending ours, and even debating whether trying to hook up with them at all is a good idea.

Ufology

To arrive at a serious discussion of those technologies and their rationales you have to wade through a lot of distracting and irrelevant junk, highlighted by the ridiculous claims of ancient astronauts by Erich von Däniken in his (yes, bestselling) *Chariots of the Gods* and its ilk. Then there's Area 51, Roswell, New Mexico, and a crashed alien spacecraft. On June 14, 1947, a rancher named W. W. "Mac" Brazel found "a large area of bright wreckage made up of rubber strips, tinfoil, and rather tough paper and sticks."[1] A DIY intergalactic spacecraft if there ever was one. And there's an ongoing history of UFOs dating from the late 1940s. The 2019 Canadian UFO survey claims Canadians see two or three UFOs *every day*. It's important to remember that many of those are the planet Venus or a hot-air balloon.*

Interest in UFOs has been revived by the US Navy's release of verbal descriptions and photos of weirdly behaving flying things, no longer called UFOs but Unidentified Aerial Phenomena (UAP).[2] These recent revelations are intriguing—no doubt about that—and while they fall far short of evidence for extraterrestrials, they're definitely a cut above ancient astronauts.

This recent version of the UFO saga began in 2004, when two US

---

* A generalization dramatically strengthened by sightings in February 2023.

Navy pilots using both radar and actual infrared video independently detected the presence of something airborne off the Pacific coast of Mexico. That something, according to their testimony, moved in a way that no aircraft known to them could move: dropping suddenly from an altitude of 50,000 feet to 100 feet, leaving no evidence of exhaust or any propulsion system, and generally behaving in an "erratic" way. Again, like nothing with which they were familiar. Could they have been mistaken? It's unlikely given that the mystery object was detected by both radar and photography. When quizzed, the pilots ruled out the possibilities of a weather balloon or birds (common explanations for such UAPs): too high to be birds, too fast to be a balloon. (Pretty cool, too: apparently every pilot who saw the indistinct video wanted to fly whatever it was.)

That was the beginning, at least of the current UFO/UAP renaissance, but not the end. In 2021, recognizing growing public interest and even pressure, the US Office of the Director of National Intelligence issued what they called a "Preliminary Assessment: Unidentified Aerial Phenomena."[3]

In it they referenced 144 encounters in a variety of circumstances; eighty involved more than one sensor, again suggesting, as above, that flaws in a single technology were not to blame. Eighteen moved in odd and unfamiliar ways. The report suggested that if and when these mysteries are resolved, they would occupy one of five categories: "airborne clutter (birds, balloon, or even airborne plastic bags), natural atmospheric phenomena (ice crystals, thermal fluctuations), US Government or US industry developmental programs, foreign adversary systems, and a catchall 'Other' bin." No specific mention of aliens, so fans of the idea that they're alien spacecraft have to be content with "Other." The US government obviously wouldn't admit if these belonged to them and are obviously most concerned with "foreign adversary systems."

That was 2021; since then an additional 366 more sightings have been reported. Of the 366 newly opened cases, 195 have been resolved with the relatively mundane explanations listed above, like balloons or plastic bags; there are still 171 that are "uncharacterized and unattributed." But no mentions of alien spacecraft.

Few conclusions can be drawn from these reports, except that a handful of observations are truly unexplainable at this time. "Unexplainable" falls pretty short of "evidence," so for the moment the idea of alien spacecraft having visited us is a hypothetical, no matter how much some want to believe it.[4]

Yes, There *Is* Some Science

Although I've already mentioned the search for extraterrestrial intelligent life, it is but one of two kinds of extraterrestrial life that are the current target of searches. The other is any kind of life at all, but predominantly microbial. There are good reasons for this focus. First, the discovery of any kind of life in the universe would be an astonishing milestone, difficult to accept for those who believe the Earth is unique, but exhilarating for many (and probably ignored by the rest). Microbes on Mars, or one of the satellites of Jupiter or Saturn, would be sensational, if not visually spectacular. And microbial life is probably the best bet for life somewhere else in the solar system. The extreme places where microbes thrive on Earth (superhot springs, rocks deep below the Earth's surface, vents at the bottom of the ocean) encourage the hope that they could withstand the even greater extremes elsewhere.

It's important to remember that any expectations we have for life elsewhere are based on our admittedly inadequate understanding of how life began on Earth. But while the exact details are still mysterious, some common and crucial elements are agreed to be necessary for life to emerge: a source of energy, water, and organic molecules.

There is water at the Martian poles even today, large amounts of it, and even evidence of a buried lake under the southern ice cap. But the heavy radiation, thin atmosphere, and low temperatures on the planet have convinced many that life—at least of the kind we know—might have flourished there in the past, but has either died out or hangs on desperately, perhaps underground. Some studies have claimed that the microbes on Earth that live at great depths in rock serve as a model for some Martian microbes, but getting deep enough into the Martian soil to find out for certain is a significant challenge. However, both the evidence for a more equable past and the familiarity of landing, traveling on, and even hovering over the Martian surface make it a prime target. But it's not the only one in our solar system.

For example, Enceladus, one of the moons of Saturn, is a fascinating second choice. It's a small moon, only 500 kilometers or 310 miles in diameter, completely covered by a layer of ice over a liquid ocean. Despite its distance from the Sun, Enceladus's ocean remains liquid because of tidal heating created by the nearby giant Saturn. The Cassini spacecraft photographed giant plumes of water spewing out into space through cracks in the ice. So there's liquid water, as well as organic molecules, carbon dioxide, and surprisingly large amounts of methane—which could point to the existence of life. Some scientists envision undersea life there could be like the undersea life clustered around the hydrothermal vents deep in the oceans on Earth.

There was even speculation that life could exist in the atmosphere of Venus, which is, generally speaking, an unlikely candidate for any kind of life because of its crushing temperatures and atmospheric pressures. A couple of years ago, a team at Cardiff University announced that they had identified the chemical phosphine in the Venusian atmosphere, a chemical that here on Earth is only produced in the lab or by living organisms. But the claim was debunked in December 2022 by a NASA team whose studies of the Venusian atmosphere found no more than tiny traces of the chemical, not nearly enough to have been produced by life.

If microbial life, past or present, is discovered in our solar system, it will reinforce the scientific story of how life began on Earth and undermine creation myths, at least those thought to be literal.* At the same time, it will strengthen support for the idea that there might be intelligent life out there. If life can start elsewhere, why couldn't it continue to evolve elsewhere, too? Obviously, finding microbes on Enceladus won't

---

* One of the greatest missed opportunities of my career was when the late Duane Gish, a prominent creationist, admitted to me that finding life on Mars would be difficult for those who believe God created the Earth and all life a few thousand years ago. Missed opportunity because he didn't say it on tape.

encourage the idea that there might be an Atlantis there, but if life can be found in two places it should be able to be found in two thousand or two million—remember, the universe is vast beyond comprehension. And if there are planets and moons that have life, countless of those would have been around much longer than we have.

So Where Are They?

Microbial life deserves the highest respect, but we have in our oh-so-human way really been devoted to the idea of *intelligent* life elsewhere. How else do you explain Percival Lowell's dogged efforts in the early 1900s to prove that there were canals dug on Mars by a dying civilization, or Iosif Shklovsky's attempt decades later to interpret odd data on the orbit of Mars's moon Phobos to argue that it was hollow and therefore an abandoned spaceship? (To be fair, it's not clear he intended this claim to be taken seriously.) I could go back much further in time, too, but the focus changed in 1961. That's when the late astronomer Frank Drake created the Drake equation, an attempt not exactly to quantify—there are just too many unknowns—but at least to get some idea of what the odds are that intelligent aliens exist somewhere in the universe. The equation is composed of a series of estimates, beginning with the assumption that, like us, intelligent aliens somewhere are living on the surface of a planet.[5]

The equation kicks off with the number of stars in the universe, because it's stars around which planets orbit. The number of stars is so incomprehensibly large that it isn't really a limiting factor. How many of these stars have planets orbiting them? When Drake invented his equation, that was an unknown number—today, astronomers have found more than five thousand planets orbiting other stars (and there is evidence that moons of such planets could harbor life as well). However, many of those planets are outside what is called the "habitable zone," that distance from a star that allows water to exist as a liquid. This does reflect a bias of a water-dependent form of life like us, but let's go with it and ignore those planets that are so far away from their star they're ice-bound and those so close that water could exist only as steam and has likely blown off. As far as anyone knows, even that number is going to be huge, as multiple searches for these so-called exoplanets continues.

So far, we're not quite halfway through the Drake equation, but the picture is a bit clouded hereafter. The next element in the equation is the number of those habitable planets that actually gave rise to life. Now we're stuck. Earth is the only planet we know of that supports life. Extrapolating from N=1, a single example, doesn't make sense scientifically. At this point if you believe we will eventually find intelligent life in the universe, your opinion is exactly that—a belief, a conviction, not a data-supported conclusion. This is why the search for microbial life in our solar system is so crucial.

You can tell we've moved fairly quickly from evidence to speculation, but let's forge ahead anyway. The next factor in the Drake equation is the number of planets that not only generated life but *intelligent* life at that. This is where I repeat: UFOs and Erich von Däniken books are not evidence of intelligent alien life. What would be? This is where we start to engage with the technologies that might come into play, but let me first bring the Drake equation to its conclusion. The sixth term is the number of intelligent civilizations that have developed advanced *communication* technologies. The seventh, no doubt a reflection on our own, is the number of such civilizations that survive long enough for us to detect them. These last two terms are unknowable, at least at the moment, although the absence of any incontrovertible evidence for intelligent aliens might say something about the last term. (Legendary Italian physicist and creator of the first nuclear reactor Enrico Fermi put it best: "Where is everybody?")*

The relevance of the Drake equation is constantly under review (after all, it's more than sixty years old), but remarkably it still stimulates thought and action. The situation, however, is frustrating because the

---

* Some argue aliens never bother visiting because Earth has such a low rating; only one star.

search for extraterrestrial intelligent life still hasn't turned up anything. There have been what looked like tantalizingly close calls, though.

I Could Have Sworn . . .

In 1967, Irish astrophysicist Jocelyn Bell Burnell's identification of a repeating signal from a distant star turned out to be the star itself, a pulsar rotating rapidly and throwing a beam of radiation across the universe. Burnell's discovery wasn't part of a search for signals from another civilization, but another, the repeating signal which was called the "WOW signal," was.[6]

In 1977, the "Big Ear" radio telescope at Ohio State University recorded one minute and twelve seconds' worth of radio wave signals from deep space that had the characteristics of the kind of signals that intelligent beings (us!) use for TV, AM/FM radio, and radar. These so-called narrow band signals are different from the typical outer-space radio waves from stars or galaxies. But while astronomers may have pinpointed the source of these signals—a star with the catchy name 2MASS 19281982-2640123—there has never been another burst of signals like WOW. It could have been a passing Earth satellite or something as yet unexplained, but if it was aliens they've been too shy to communicate further.

Just a few months ago, Chinese scientists announced "several cases of possible technological traces and extraterrestrial civilizations from outside the Earth."[7] This one had a quick turnaround: two days later, it was concluded that the signals were generated on Earth. To avoid earthly contamination of such signals, we would need radio telescopes on the far side of the Moon or somewhere in a distant orbit around the Earth.

We Are Advertising the Earth

These stories share a common thread or at least hope: there are civilizations out there that would choose to broadcast signals that other intelligences would understand as artificial, not natural, and would contain actual information. Ironically, we have been doing exactly the opposite since the early 1900s, because radio and television broadcasts leak out and travel from Earth at the speed of light, meaning that traces of those early shows have already encountered an estimated seventy-five star systems, four of which have planets. But our reach isn't infinite; the farther away from Earth the broadcasts spread, the weaker they are, and a very distant alien civilization wanting to watch *I Love Lucy* (a sure sign of high intelligence) would need to have built a giant radio telescope.

There is concern that maybe we should stay quiet and not advertise ourselves, given that we are a technologically young civilization and there could be much more advanced versions out there that aren't necessarily friendly. Unfortunately, it's too late and not just because of radio and TV leakage. If an advanced civilization had been keeping an eye on the chemistry of the Earth's atmosphere, they would know that dramatic changes have been happening for the last few thousand years, especially changes in the chemistry of the atmosphere wrought by the Industrial Revolution.

Should we be sending more targeted messages? Besides the caution expressed above, time is an issue. Targeting a star fifty light-years away means there won't be an answer for at least a hundred years (and if that civilization has bureaucrats, much longer); that challenges the human attention span.*

The search for clues in radio waves traversing the universe is sure to continue, but it's worth questioning some long-held assumptions about aliens. Like, why do we think they're so much like us?

What Should We Be Looking For?

You don't have to be a fan of science fiction to know that almost all aliens that have populated tales of the future have much in common with us:

---

* A shout-out here to the many builders of the great Gothic cathedrals who never saw the completion of their projects.

two-legged, four-limbed creatures with heads (often oversized) more or less like ours (that physicality used to be necessary when human bipeds were hidden inside the costume, but not so much now when they can simply be created by CGI). And these humanoids build settlements on planetary surfaces, just like us, or even advanced space station–like structures—like us. The anthropomania of it all extends even to the chemistry we seek: "Molecule X should appear in the atmosphere of an exoplanet if there's an industrial civilization there." Yes, there are good chemical reasons that carbon, for instance, being a versatile molecule, might well serve as the basis for life elsewhere, but given that we have very little experience with the fine details of extraterrestrial chemistry, there's really no reason to close our minds to other possibilities. Of course, the problem is that it's difficult to imagine something you can't imagine.

The English astronomer Fred Hoyle did in his 1957 novel *The Black Cloud*. The "cloud" was a massive accumulation of dust and gas that wandered through the universe living off the energy of stellar radiation. Upon entering our solar system, it threatened to move close enough to block the Sun from the Earth, which would have been the end of us. The cloud chose to move on, though, and I use the word *chose* deliberately—it had an intelligence greater than human.

It's worth mentioning that Hoyle was a prominent astronomer in his day. Another, even better known, was Carl Sagan and he, too, threw out a radical model for life elsewhere, arguing there could be living organisms floating in the atmosphere of Jupiter, surviving by soaking up solar radiation and/or a variety of organic molecules.[8] Sagan didn't claim intelligence for these extraterrestrial jellyfish, but they do represent a very unearthlike model of alien life.

I'm not saying Hoyle's black cloud is necessarily a good candidate for an intelligent life-form, but it does force us to ask ourselves, could the hypothetical ones out there be just as foreign to our concept of life? And if they are, how should we try to communicate with them, or, if we judge that to be a rash step, even detect them?

Some scientists offer the counterargument that the humanoid model is actually a good one for extraterrestrials. Zoologist and astrobiologist Arik Kershenbaum at Cambridge University argues that while Earth is the only example of life that we know, there are principles operating here that have driven life in certain directions.[9] Bilateral symmetry, the right side–left side organization of animals, has an incredibly long evolutionary history, beginning in the early oceans and guiding the development of fish, insects, birds, and mammals. Such symmetry apparently enhances the ability to maintain a straight line when locomoting and also to make abrupt turns. Perhaps the same gains in efficiency of movement are a good thing even on other planets.

Flight evolved very differently in bats, birds, and insects, but while the structure of their wings is radically different, the common element is that flight is a good way to get around on a planet with an atmosphere. It's more versatile than that: rays are examples of "flying" in the much denser medium of water.

Organisms living on land, everything from insects, amphibians, and reptiles to mammals, evolved legs for movement. Would legs be common on some Earthlike exoplanet? The structure of legs on Earth is constrained by the physical environment, especially gravity. Insects have incredibly skinny legs; elephants have legs like columns. Each is determined by the ratio of weight to volume that needs to be supported. These factors are driven by gravity, which we know is universal. Kershenbaum argues that it wouldn't be surprising to see that some form of natural selection is universal and would prompt a familiar sort of evolution of life in response to the physical and chemical environments elsewhere in the universe. Whether intelligence would eventually appear is beyond unknown.

Can We Find Them Even If They're Not Calling?

How might we detect alien intelligences?

Geoengineering scholar Andrew Lockley and climate scientist Daniele Visioni published a paper in the *International Journal of Astrobiology* called "Detection of *Pre-Industrial* Societies on Exoplanets" (emphasis mine).[10] In it, they acknowledge the bias underlying the search for signs of intelligence based on our only example—Earth. But they also point out that it is just impractical to search for nonhumanlike beings like "liquid-methane dwelling octopodes with a millennial lifecycle" or "ultra-fast ultra-dense lifeforms living on neutron stars." Hard to disagree with that.

They further point out that given that we have been technological, which they define as "radio-capable," for only about a hundred years, but have been agriculturalists for ten thousand years, it makes sense to look for pre-industrial activity on other planets, assuming of course they are following a similar course as ours. But what would such activity look like given that we're looking at planets light-years away?

The authors' conclusions are that some pre-industrial changes might be seen from astronomical distances, but there are limitations on each: if a civilization makes large-scale changes either to land or oceans, those might be detectable as alterations in reflectivity or the chemical composition of the atmosphere, but either would require long-term observations and again they are predicated on the hypothetical civilization following roughly the same path as we have on Earth. It's also hard enough to detect *any* changes over such vast distances, but it's even more difficult because pre-industrial societies would presumably be making much less of a mark on their world than their post-industrial counterparts. Given those challenges, it seems unlikely that substantial efforts to detect such changes are on the table, at least at the moment.

Detecting post-industrial changes is more promising. This is the field of "technosignatures," visual or chemical changes similar to those that have happened on Earth since the nineteenth century. City lights, clearly visible from Earth's orbit, would be one; nitrogen dioxide or chlorofluorocarbons accumulating in the atmosphere might be another. Nitrogen dioxide is generated naturally from volcanoes or lightning strikes, but industrial production of the gas is far greater and could be significant.

Besides city lights (would they have cities?), heat islands produced by high concentrations of industrial activity might be detected by infrared telescopes.

"I Could Have Sworn 2.0 . . ."

If a distant civilization constructed megastructures, either on a planetary surface or in orbit, and those either absorbed or reflected starlight, the contrast between the structures and the natural surface of the planet would be a giveaway. In fact, there was a flurry of interest around exactly that a few years ago when a star that displayed wild fluctuations in its light was detected by earthbound telescopes. Minor changes like that are common when an orbiting planet crosses in front of a star, but such interruptions are slight, predictable, and temporary. In this case, the light from the star was dramatically and irregularly interrupted, puzzling astronomers. Despite its prosaic name—KIC 8462852—it was deemed to be "the most mysterious star in our galaxy."* The star, or rather the fluctuations in its light, was discovered in 2015, but even today there's no completely accepted mechanism for what's going on. It's not the presence of another star (the changes in light would be regular) and the star isn't pulsating; it's not a swarm of comets, and even the most recent speculation—that it is a disintegrating moon that has somehow wandered away from its home planet and is now on its own (the charmingly named "ploonet")— is as uncertain as all the others. At the moment, the dimming is thought to be caused by gigantic, irregular clouds of dust.[11]

There was one suggestion that was even more out there: the idea that the interruptions of light from KIC 8462852 were caused by some kind of megastructure built by aliens orbiting the star. Curiously, this was an idea floated in 1937 by British science writer Olaf Stapledon but made popular by physicist Freeman Dyson in his 1960 paper "Search for Artificial Stellar Sources of Infrared Radiation."[12] Dyson proposed that a planetary civilization faced with an impending energy crisis might build a network of solar collectors around their star to establish a permanent source of energy. Surrounding the star would enable them to collect many times

---

* Lest you think astronomers are incapable of coming up with cool names, KIC 8462852 was quickly dubbed "Tabby's Star" after astronomer Tabetha Boyajian, who discovered it, and is now more formally known as Boyajian's Star.

more energy than confining solar collectors to a planetary surface. The extreme version, called a "Dyson Sphere," would completely encircle a star (although holding such a sphere together would probably be an insoluble challenge). Could KIC 8462852, "Tabby's Star," be explained by the ongoing construction of such Dyson objects?

Who knows? Since 2015, other stars exhibiting these strange dimmings have been discovered and no one is arguing that they are all aliens. And at least one of them is pretty dramatic: something has dimmed a giant star by more than 90 percent for a couple of hundred days, a "something" that was first described as a large, elongated object with a "hard edge." Nothing more is known.

Tabby's Star and these others are perfect examples of the combination of intrigue and skepticism that surrounds the search for extraterrestrial intelligent life. If you accept that many or at least some of the processes that occurred on Earth to generate life are common throughout the universe (and there's a host of elements and molecules that exist both here and in space), then other civilizations, or at least other organisms, make sense. Our ability to detect them has a long way to go before we can say much about their possible existence. But unless they come to visit us (and some have suggested that the answer to Fermi's question "Where is everybody?" is that they detect no sign of intelligence on Earth), we will use technology to find them, and finding them would be the ultimate discovery.

PART IV

Changed by Computing

CHAPTER 13

Robots

Our mental images of robots are deep-rooted, informed much more by literature, movies, TV series, and our own wild imaginings than by the technologies themselves. In fiction, technology is no barrier—robots can be anything you want. So the clunky, box-headed robots of movies past have become sleek and compellingly humanoid. But their development isn't limited to being more like us. In fact, the diversity of robots is expanding dramatically, driven particularly by their purpose. There are now robots for everything from unloading boxes in a warehouse to exploring the depths of the ocean, or, as you'll see in this chapter, flying—and landing—like a bird.

There's no mystery to our fascination with robots: they are like us, but they're not. In them we get to "meet the other," like our ancestors met the Neanderthals. And like the Neanderthals perhaps should have realized, it's important to be wary of them. They might be dangerous. But whatever image you're holding in your head right now, the robot in your mind is always a blend of fiction and nonfiction: the technology of robotics, the artistic imagination, together.

Students of the history of robots usually identify Mary Shelley's *Frankenstein* as one of the key works that shaped our views. *Frankenstein* had a curious origin: In 1816, Mary Shelley and her new husband, poet Percy Bysshe Shelley, were guests of the poet Lord Byron and his physician at an Italian villa. One evening after swapping ghost tales, each agreed to write one of their own. Mary Shelley finished hers in 1817, and it was finally published anonymously on New Year's Day 1818, in London. Shelley was twenty. Even though Frankenstein's monster wasn't metallic or wooden but living flesh created from the nonliving, the idea of an eight-foot-tall humanoid creature wandering the earth killing people is just a short step away from the Terminator, who at least had humanlike skin (the T-800 version).

But . . . *Real* Robots, Too

Shelley's *Frankenstein* was not the first time—by far—that a synthetic humanoid had been imagined. Not just imagined but actually built. While I'm not sure you could call this a robot, a dictator named Nabis, who ruled Sparta around 200 BCE, was extremely cruel and barbaric, exceeded in these qualities only by his wife, Apega.[1] Together they tortured and killed people for money and jewelry, and according to the historian Polybius, Nabis had a *mechanical replica* of his wife made, which was said to "resemble his wife with extraordinary fidelity." Why would he do that?

If he was unable to persuade a guest to hand over his valuables, Nabis would introduce his mechanical wife, and standing behind "her," have her extend her arms to embrace the guest, with the one complication that her body was covered with sharp iron spikes that, as she hugged him closer—and closer—would kill him. So it was an artificial being in a sense, but incapable of movement herself, and really more a macabre torture instrument than a robot.

But millennia ago there were even more elaborate automatons built for display and wonder, not murder. In 279–78 BCE, the Greek ruler of Egypt, Ptolemy II, orchestrated a procession celebrating the Greek god Dionysus. In addition to chariots, giraffes, lions, elephants, and ostriches, the procession featured hordes of singers and musicians and, most spectacular of all, a roughly four-and-a-half-meter (15-foot) statue of Nysa, Dionysus's nursemaid. Seated on the ancient version of a float, this giant automaton would periodically stand up, pour milk from a bowl, and sit down again.[2]

It all sounds kind of unbelievable—a statuesque 2,300-year-old robot? But in the last few years mechanical engineers have outlined a scheme of gears, cams, and sprockets that could explain how, as Nysa's vehicle rolled along, she was able to stand, pour milk, and sit down again. The mechanisms necessary were either known to the ancient world or can reasonably be assumed to have been known.* The clever inventor of this object has not yet been identified. There are suspects, but that's all.

Alexandria, Egypt, was at that time a dramatic and creative place. But the urge to create human replicas out of inanimate material is probably much older and has never wavered since. Medieval times, the Renaissance, and of course the Industrial Revolution saw the rise of new mechanisms, new tricks: a duck that ate and excreted; a bird that extended its wings in alarm when a snake approached; even puppet theaters.

Then Along Came Asimov

Robotics continued to advance, finally acquiring a name in the twentieth century. Most references claim that the word *robot* was invented in the early 1920s by Czech author Karel Čapek in his play *R.U.R.* (which stands for *Rossum's Universal Robots*).[3] Apparently, Čapek had the idea for the play and consulted his artist brother, who spat out the word *robot* (apparently, he had a paintbrush clenched in his teeth at the time). *Robotics* itself didn't appear until Isaac Asimov coined the term in 1941 in his science fiction short story "Liar!", in *Astounding Science Fiction* magazine.[4]

---

* A good reason for confidence in the technology of the time is the Antikythera mechanism, a stupefyingly complex bread-box-sized device found in a shipwreck off the coast of Greece about a hundred years ago. It dates back two thousand years, and its gears and wheels predict eclipses of the Sun and Moon and track the movements of the five planets that were known at the time. Such intricacies don't appear again for another one thousand five hundred years.

Throughout the twentieth century, one technological challenge in robotics after another was solved, but actual progress was nothing like the evolution of fictional robots. Robots in popular culture took off in the twentieth century.

In 1942, the year after Asimov invented the word *robotics*, he came up with the truly legendary "Three Laws of Robotics."[5] For those who might have missed them:

1. A robot may not injure a human being or, through inaction, allow a human being to come to harm.

2. A robot must obey the orders given it by human beings except where such orders would conflict with the First Law.

3. A robot must protect its own existence as long as such protection does not conflict with the First or Second Laws.*

Should We Fear Them?

Any science fiction reader knows the three laws, but those with a serious interest in robots and how they'll develop in the future point out that the laws have serious shortcomings and so are inadequate to safeguard us as robots evolve. In the journal *Frontier Robotics and AI*, Christoph Salge and Daniel Polani of the University of Hertfordshire argue it would be hard to make the laws workable.[6] (It should be said that most who comment on the three laws suspect that Asimov himself knew they wouldn't work, pointing out that he wrote several subsequent stories about robots that proved that.) One of the issues is language. The problem arises in the first law: in Salge and Polani's view, the word *harm* is difficult to make

---

* Asimov later added a fourth law, which, because it was supposed to be the most crucial and therefore should appear at the top of the list, was called the "zeroth law": A robot may not injure *humanity*, or, by inaction, allow humanity to come to harm. It shares the same shortcomings with the others.

meaningful to a robot, at least a current robot. And if you're dealing with harm in, say, the legal sense, it's ambiguous enough already. Because the best decisions usually depend on the specific situation, these scientists feel that robots should instead be "empowered," flexible enough to deploy different strategies depending on the situation. In this context, empowerment means letting robots perfect ways of achieving the goals of the three laws, absent the hurdles presented by them.

Empowerment would also apply to humans, so that robots would be obliged to maintain the empowerment of a human. Locking a human in a room would not be permitted, unless of course there was a fire and locking them behind a fireproof door might actually improve their long-term empowerment.

Another example Salge and Polani offer is the rule "Don't push humans." Most of the time it makes sense, but if an out-of-control car is about to mount the curb and strike someone, pushing them out of the way is the correct action. Only a robot whose rules allow flexible behavior would be capable of sizing up the situation and overriding its default rule.

Swedish-born philosopher Nick Bostrom has expanded on the problematic term *harm*, wondering how a robot could "balance a large risk of harm to a few people versus a small risk of many humans being harmed." He, too, is concerned about definitions: How does a robot differentiate between harm causing physical pain and the harm of injustice?[7]

In its much more recent Principles of Robotics, the United Kingdom's Engineering and Physical Sciences Research Council (EPSRC) argues that Asimov's Laws should more appropriately be aimed at people—the designers and makers of the robots.[8]

At the moment, such attempts to prevent a robot from wreaking havoc on humans are a greater concern for artificial intelligence researchers, for fear that superintelligent AI, if and when it develops, simply won't care about our long-term welfare. Whether malevolent AI would even bother with robots isn't clear—they need not be the actual instruments of AI's (hypothesized) attempts to take over (more on this in Chapter 15). The "threat" from robots has always been seen as physical, but of course in the end advanced robots will be furnished with AI anyway, so one way or the other the problem must be solved.

Robots of the Twenty-First Century

Has the technological development of actual robots kept pace with the robots in popular culture? Humanoid robots have of course been the stars of film, television, and literature, and as depicted there, would be extremely difficult to distinguish from a human being (see replicants in *Blade Runner*). In the real world, there have been big strides forward in the development of humanoid robots, the kind that will likely take on social companionship roles, but at the moment they are only part of the picture. Diverse robots are where much of the action is today.

The American company Boston Dynamics is famous for the robots depicted in their videos, many of which are four-legged. They resemble animals—of some kind—and move with uncanny animal-like fluidity. They are definitely not humanoid, although Boston Dynamics has built those, too.[9] Why should they be? Think of the robots on the car assembly lines that execute their limited repertoire of maneuvers over and over again. Do they look like any sort of living thing? No. They're intended to be powerful, effective, and economical, not attractive. And so robotics has flourished, just not in the way that science fiction might have anticipated or even been interested in.

A robot that caught my eye recently is called SNAG: Stereotyped Nature-Inspired Aerial Grasper.[10] This is a birdlike robot, but the nature-inspired part is not birdlike flight, but landing, especially the tricky challenge of landing on a branch. Prior to SNAG there had been many attempts to create robots that could mimic bird landings; all contributed some pieces of the puzzle, but none was particularly effective. The scientists behind SNAG, engineers at both Stanford University and the University of Groningen in the Netherlands, built their bird-claw robot on principles derived directly from the mechanisms of actual bird claws. As they point out, a bird landing on a branch is really making a controlled collision, and their robot mimics both the bird's approach, dialing down velocity as it nears the landing, and impact, when the legs collapse and the energy from flight is transformed into energy to grasp the branch. Then small bodily adjustments position the bird securely. They went to the trouble of using materials to build their landing apparatus that are analogous to the ligaments, tendons, muscles, and feet of real birds. The flight of their system is provided by a quadcopter. In their videos, SNAG lands successfully and immediately flattens itself on a branch. Having modeled the shape and size of some of the components on peregrine falcons, the scientists also tested SNAG's ability to catch objects thrown at it, almost like plucking a pigeon out of the air.

They suggest that robots with flight ability and their landing gear could be ideal forest ecosystem monitors, especially because they are not required to stay airborne all the time. Imagine a forest full of robots flying, landing, and taking off. Or the robot carrying your latest Amazon package pausing to perch in the big spruce out front before delivery.

But a robot that lands like a bird is only one small example of the hordes of modern robots. In agriculture, robots to rid crops of pests, gather honey from beehives, and pick fruit are either already in use or are being developed. True to its name, "Slothbot" glides very slowly along wires suspended in the forest canopy to monitor temperature and carbon dioxide levels. "Slugbot"? Yes, it picks slugs out of crops and disposes of them.

There are even robots under development to pollinate plants in an effort to cope with the variety of challenges that have faced honeybees in the last decades. None of these, ranging from vehicles with robot arms to transfer pollen from flower to flower to flying bee-like pollinators, are in general use yet. Nonetheless the TV series *Black Mirror* capitalized on the idea of the insect version in its episode "Hatred in the Nation," in which artificial pollinators were reprogrammed as tiny winged assassins.

Who Cares If They're Humanoid?

When we picture humanoid robots, we focus on appearance and movement, especially subtleties like facial expressions. It's simply not possible to adapt the ubiquitous assembly-line robot to that end because it's made of the wrong stuff. Such robots are composed of rigid materials like steel and aluminum to be able to lift great weights with precision and are usually restricted to stereotyped back-and-forth and up-and-down movements, their utility being that they can extend these tasks unfailingly 24/7. The movements are restricted to those necessary to enable their function. Such robots are usually not capable of adapting to new routines, nor are they designed to. This rigidity is a bonus on the factory floor but not applicable to myriad other settings. This has opened the door to designing and manufacturing "soft" robots.

Substitute soft, pliable limbs and a rotund—even pudgy—body for the rigidly metallic and you are getting closer to a soft robot. Actually, the octopus is the biological model (although an elephant's trunk is suitable, too)—roboticist Cecilia Laschi was inspired by the octopus.[11] The octopus has no bones but is capable of amazing dexterity by virtue of its eight tentacles (much more adaptable than arms), each one controlling its own movements by virtue of a mini-brain—much of the octopus nervous system has been off-loaded to the tentacles. You're not going to see a robot like this on any assembly line, but you might see it on the bottom of the ocean.

It should come as no surprise that nature should be the inspiration for soft robots. The late biologist Steven Vogel made a career of understanding and explaining how forces shaped structures in nature. He pointed out that a cat's or dog's ear rotates to point in the direction of an interesting sound, as does a satellite dish with respect to a signal. But the ear has no stiff parts, just skin, muscle, and cartilage, all of which can bend. He argued that our experience with building has given us great skill in working with stiff materials, like metal and wood, but much less with soft. Soft robotics might change all that.

Recently, a consortium of Chinese scientists created a soft-body robot that looks a little like a fish with a long tail and broad fins extending out to the sides. Its "muscles" are driven electrically, and its body is mostly made of deformable silicone. This is a seafaring robot, and the most astounding thing about it is that it has been taken to the deepest part of the ocean, the Mariana Trench, 11,000 meters or nearly 7 miles down.[12] While in the first test run it was attached to another device, it was able to flap its fins and its electronics survived the enormous pressure. While the experiment immediately suggested that this robot might be capable of exploring the depths of the ocean in ways never before possible, or even imagined, it also suggests a future for soft robotics in humanoid robots.

The traditional assembly-line robot is unsuited for any kind of human contact—in fact, they are often enclosed in a cage to prevent that very thing. They are huge, heavy, and powerful. Soft robots are the antithesis of that, suggesting that they might be a more suitable model for interacting with humans. Of course, humanoid soft robots are not going to look like the undersea explorer, nor will they be built in exactly the same way. Just as we are a combination of soft tissues, bones, and cartilage representing a range of hardness and rigidity, it's likely that humanoid soft robots will be as well, to preserve their shape-changing adaptability while adding strength and speed.

One of the beauties of these soft materials, which are often sophisticated cousins of rubber, is that they can extend or contract in response to electrical impulses, which in turn are generated within the structures. Like muscles, some of these materials, called "shape-memory alloys," are able to change shape and then regain their original form. The result is a self-contained, flexible, and multipurpose limb, lever, or gripper, strong enough to lift and gentle enough to grasp an egg without crushing it.

Once advanced enough to be deployed, soft robots are likely to serve a variety of purposes, from emergency response to exploring hard-to-access places, like the Mariana Trench explorer, to more pliable, responsive, and comfortable exoskeletons and flexible tools for surgery—even robots that interact with humans.

Soft robots are still far from being ready for human contact, although the technological developments needed for autonomous control, power supply, and appropriate building materials are under way. But technology is only part of the picture. Designing robots that will interact with humans brings us face-to-face with the human acceptance side of the equation. The issues are diverse, but one of the most fascinating connects directly to our millions of years of evolution. The importance of social life has meant that we have an extremely finely tuned sense of what seems natural, and what is just a little off and a little disconcerting. A slight hesitation in the way a robot's legs move, the tilt of its stance, and above all, the not-quite-human expressions on its face, both while talking and listening, are all giveaways.

One example of the trickiness of building true-to-life humanoids is a psychological phenomenon called the "uncanny valley." For fifty years, the uncanny valley has occupied the attention of psychologists, roboticists, and animators, yet it's not completely certain the uncanny valley is a real thing. Some experiments support it; some don't.[13]

A typical experiment has volunteers rating their comfort or trust in a set of faces that gradually transform from caricature to natural human. The uncanny valley predicts that at a crucial point, just before the face becomes unmistakably human, it disturbs and even distresses those looking at it. In those experiments that have confirmed the effect, it truly is a valley, rapidly descending from comfort to discomfort as the face nears humanness, then rising abruptly again to be completely comfortable with the fully human face.

It's possible you've experienced something like this yourself, especially if you saw the movies *The Polar Express* or the 2007 production of *Beowulf*.[14] Critics and viewers alike felt a certain unease from seeing not-quite-Tom-Hanks or Angelina-Jolie-almost-nude-but-wait-that's-CGI. Comments like "They look human but not human enough" and "The characters have lifeless eyes, black eyes, like dolls' eyes" were common on social media. Some humanoid robot designers have even gone to the lengths of not making their robots look too human, by omitting hair and making them grayish rather than flesh-colored. However, the recent proliferation of absolutely realistic, completely computer-generated facial expressions suggests that at least when it comes to faces, the uncanny valley may soon fade in importance—if it hasn't already.

When Robots Become Personable

One of the most celebrated humanoid robots is Sophia, created by Hanson Robotics.[15] Designed to interact with and educate humans about the future of robotics, she is actually a citizen of Saudi Arabia (which, of course, raises the question of what rights—if any—she would enjoy there as a female robot) and is equipped with artificial intelligence so she can, at least theoretically, learn and become more humanlike with time. YouTube videos of Sophia reveal that while she possesses an array of facial expressions, she still has a long way to go before she would be mistaken for human. Yet she, and others like her, suggest the direction humanoid robots are going.

As I've said, the technology is only one side of the coin; the human response is the other. And already, before humanoid robots are anywhere near perfection, some unexpected warning flags are being raised about what seems at first glance to be the most innocent robots designed for inarguably laudable goals—small robotic pets to alleviate older people's loneliness. There's no doubt that loneliness is a painful situation that can shorten life. That was clear even before COVID. Several studies have shown that lonely people can be cheered up by, even become strongly attached to, small robotic cats and dogs—faux animals that do little more than make a few noises and some simple movements. Harmless, right?* Beneficial, actually. But there are dissenters. One of the most prominent is philosopher Robert Sparrow at Monash University in Australia.[16] He argues that by giving such robot "pets" to people, we are either deceiving them or asking them to participate in their own self-deception, which he finds immoral. While he admits robot pets are easier to care for than the real thing, he doubts that they provide the same kind of intimacy and

---

* Although far from humanoid, the MIT Media Lab's "Jibo," a squat little thing that looks more comfortable on a kitchen counter, is intended to be a human companion. In Jibo's own words, "I just want to be there for you and help you when you need me most. You are the reason I'm here. After all, if there weren't humans, we wouldn't need robots." www.jibo.com.

comfort the real ones do. He extends this argument to foresee that the introduction of humanoid robots in the care of the elderly, a practice that many have argued would be beneficial, would in a similar way reduce the respect and recognition central to such care.

Whether you buy that argument or not, an extension of it offered by others is that having robots involved in elder care is an example of the human tendency to look for technological solutions to situations that actually need to be reworked from the bottom up. Imagine this possibility: social care robots being adopted as the default care providers, resulting in requests for real human care being viewed as a luxury or indulgence—which of course would have to be paid for. Impossible? I doubt it.

Some researchers are already imagining some very different issues that might arise. If "Hitchbot" is any example, the robots might need much more protection from us than we need from them. Hitchbot was a Canadian invention, a hitchhiking robot made from a beer cooler bucket and pool noodles that stood by the side of the road waiting for rides. It made it from Halifax, Nova Scotia, to Victoria, British Columbia, without incident, a journey of 6,000 kilometers or 3,728 miles. When a similar trip was initiated in the United States, Hitchbot lasted a mere two weeks before being cruelly dismantled—actually decapitated—in Philadelphia, the City of Brotherly Love. Admittedly, it wasn't a humanoid robot, but might people be equally hostile to robots that resemble them?[17]

Christoph Bartneck and Merel Keijsers at the University of Canterbury, New Zealand, explored this issue further in a paper titled "The Morality of Abusing a Robot," in the *Journal of Behavioral Robotics*.[18] They set up a clever experiment where volunteers witnessed videos showing a sequence of abusive behaviors toward a human and a robot. The trick was that the robot was actually a CGI representation of the same human—same movements, same set of abusive actions, like being kicked, obstructed with a hockey stick, or even shot with a gun (obviously fake). After fourteen such instances of bullying, the victim, whether human or "robot," retaliated twice, including hitting and kicking the bullying human. Interesting results: those watching the video judged the abuse of the robot to be as immoral as abusing the human. It wasn't clear whether they equated the two because they believe robots merit the same treatment as humans or whether they thought abuse was immoral, period, regardless of the target.

Oddly, they considered the robot's fighting back to be less acceptable than the human's. Digging deep, the researchers found this discrepancy was due to the viewers thinking the robot had been more abusive in its response. But remember, the so-called robot victim was a CGI version of the human victim, so both robot and human retaliation on the video were *literally* identical.

The authors wondered if this misreading of the robot's actions was due to the omnipresent fictional depiction of killer robots. Preconceptions certainly play an important role, especially the idea that robots, no matter how humanoid they might appear, are not human. Even Sophia, though she has attained "personhood," would not be assumed to have full human rights by most people. Neither humanlike movement nor apparently spontaneous conversation is enough to convince most of us that robots are thinking or are conscious. That's today, but as humanoid robots become more like us, surely attitudes will be forced to change. In fact, the physical embodiment of future robots will be much less important that the brains behind them. That's where artificial intelligence, AI, of a quality not yet achieved, will play the most important role.

CHAPTER 14

AI Today

Artificial intelligence, AI. It is so common it's practically unnecessary to list the ways it is already being used. Instead, I'll just apply what is said about spiders: there's likely a device using AI within a few meters of you right now. I will admit there's a little controversy here. I've seen claims that there's a spider within a meter of anyone *at all times*, but obviously that is more likely to be true in a forest than in your living room, but still . . . I'm guessing "a few meters" is more often correct than not. Bottom line: AI is everywhere.

It is busy calling Uber, listening or viewing online music or ads, banking online, planning a route using Google Maps, monitoring your home security, detecting if you suddenly fall, using face recognition to unlock your phone, configuring your social media feed, and being provided with programs "you might like" by your online streaming service. But that's not even the half of it.

"Dave, stop . . . Stop, will you? Stop, Dave . . . my mind is going . . . I can feel it . . . there's no question about it . . . I'm afraid, I'm afraid, Dave . . ."

That is the voice of HAL 9000, the computer on board the spaceship in Arthur C. Clarke and Stanley Kubrick's 1968 film, *2001: A Space Odyssey*. That HAL had a voice isn't at all that remarkable (Hey, Siri?).

That Hal could plead for "his" life is. HAL appeared to have a complete set of human emotions, a capacity that no version of artificial intelligence has today. The movie is more than a half century old, but even so, HAL touches on some of today's AI controversies.

First, let's acknowledge that HAL was vastly more advanced than any AI program today: he was self-aware and emotional, two qualities that AI designers aren't even sure will be possible—or necessary.

For all of HAL's smarts, though, astronaut Dave Bowman was able to disable him by simply throwing the right switches. Many experts would argue that AI as smart as HAL would anticipate that possibility and would have taken steps to ensure he couldn't pull the plug. Ever.

More important, *2001* got it right: it isn't hostile robots that pose an existential threat to humans. HAL was no robot: he wasn't mobile, he had no limbs; he was just a console stuffed with electronics. That in itself was significant: people are interested in people, which is why most Hollywood representations of artificial intelligence are housed in humanlike bodies. But, like HAL, they needn't be, which is why I separated the previous chapter on robots from this one: while inevitably some robots, maybe *all* humanoid robots, will eventually be equipped with AI, AI doesn't need to be equipped with robots, although of course it could deploy them if it chose to.

The existence of a malevolent AI isn't unique to *2001* (see *Terminator's* Skynet), but it is science fiction. There is no AI that is malevolent, or benevolent, or anywhere on that spectrum. AI as it exists today, while in some cases is very, very good at what it does, is nowhere near what researchers would call artificial *general* intelligence, that is, the ability to be as smart as humans in everything that humans do. So remote is AI today from that label "general intelligence" that you could make the case that we will *never* see such a machine. Or you can argue for its inevitability.*

---

* In 1968, Arthur C. Clarke predicted, "Many scientists think that in the next century we will have machines that are more intelligent than us. And of course this might be one of the great divides in history."

Babbage, Lovelace, and . . . Jefferson?

Nearly two hundred years ago English mathematician Charles Babbage invented, at least on paper, early computerlike machines, which he named the Difference Engine and the Analytical Engine. Although Babbage never managed to build either, they were quite different: the Difference Engine was really only an advanced calculator, but the Analytical Engine was much closer to being a true computer. But it wasn't Babbage who realized the enormous potential of his inventions; it was his protégée, Ada, Countess of Lovelace.*

There is still debate as to whether she should be called "the first computer programmer," but regardless, her insights were profound. Ada Lovelace understood that computers wouldn't be limited to mathematics, as Babbage thought. She saw in the use of punch cards that "the Analytical Engine weaves algebraic patterns just as the Jacquard Loom weaves flowers and leaves" and suggested that machines like the Analytical Engine could create patterns, not just of numbers, but of musical and artistic symbols. A measure of her depth of knowledge: when asked to translate a description of the machine from French to English, she did and added some of her own notes—twenty thousand words' worth, more than twice what was in the original article.[1]

Lovelace's prescience is all the more amazing given that Babbage's engines were never actually built (at least not until late in the twentieth century, and even then only the earlier version, the Difference Engine, was completed).

The shortage of appropriate materials and manufacturing techniques confined Babbage's inventions to the plans and not much more, and it wasn't until after World War II that actual computers began to be built, bringing speculations about their future with them.

In 1949, Dr. Geoffrey Jefferson, an elite neurosurgeon at the University of Manchester, wrote an article called "The Mind of Mechanical Man" in

---

* Strange coincidence: Lovelace was the only legitimate daughter of the poet Lord Byron, at whose Italian villa Mary Shelley was staying when she began to write *Frankenstein*.

the *British Medical Journal*.[2] He exhaustively detailed the similarities and differences between the brain and a computer (as they were understood at the time), convinced that there was no way a machine could ever even be a shadow of the human brain. He proclaimed, "Not until a machine can write a sonnet or compose a concerto because of thoughts and emotions felt, and not by the chance fall of symbols, could we agree that machine equals brain—that is, not only write it but know that it had written it."* In effect, Jefferson started with the brain, reached toward the machine, and found an unbridgeable chasm between them. Given the state of computers then, even a "chance fall of symbols" would be pretty unlikely to write the sonnet, let alone be aware of it. At almost exactly the same time, the brilliant English computer scientist, mathematician, cryptographer, and biologist Alan Turing attempted the same journey from the other side and, sonnets or not, found the crossing disturbingly easy.

In a 1951 speech in Manchester called "Intelligent Machinery: A Heretical Theory," Turing said, "[I]t seems probable that once the machine thinking method had started, it would not take long to outstrip our feeble powers. There would be no question of the machines dying, and they would be able to converse with each other to sharpen their wits. At some stage therefore we should have to expect the machines to take control. . . ."[3]

Seventy years later, such sharp differences in opinion still exist about the future capabilities of machine thinking (artificial intelligence) and its potential threat, even though we are much closer to having intelligent machines. Sometimes researchers with opposing opinions can be working on artificial intelligence in the same research institute, practically side by side.

If AI is nowhere near being capable of artificial general intelligence, where does it stand right now?

---

* As cynics have said, apparently no one thought to ask Dr. Jefferson if *he* could write a sonnet.

Ubiquitous AI

Most of us use AI. Ninety-eight percent of iPhone owners use Siri; 96 percent of Android owners use Google Assistant. But using AI and realizing that's what you're doing are apparently two very different things. Eighty percent of us use it in some way, but only 30 percent are aware of it—nearly three-quarters of the population doesn't realize they're surrounded by AI. I'd certainly bet that 75 percent of us have no idea how sophisticated even the most everyday devices are. For instance, to answer a question, Siri (Speech Interpretation and Recognition Interface) must be able to make sense of the acoustic data of your voice (sampled 48,000 times every second); determine exactly what it is you're asking; use vast memory banks to find the words, phrases, and places appropriate to the question; then use them to compose and articulate a reasonable answer. It's the technology's speed and flexibility that is pretty impressive.

And that's just Siri.* By comparison, any self-driving car or truck (and collectively they've already driven tens of thousands of kilometers) has to process a lot more information with much more riding on its accuracy. And the vehicles are developing quickly: there are already cars on the road that sense a collision is about to happen and decide whether it's worth swerving or not. Then they do it—or not.

I changed this scenario slightly once and asked a group of AI researchers: "Will the self-driving car be able to tell, here in Alberta, the difference between a deer and a moose running onto the road?" The question is crucial because it's generally accepted that you shouldn't swerve for the deer (sadly for the deer), because the collision is usually pretty risk-free for the passengers in the vehicle. A moose is different; it's a more unpredictable animal and much more massive, with long legs that ensure in a collision that the body clears the hood and hits the windshield. While swerving at speed is dangerous, hitting the moose would

---

* My TV Siri might not be sentient but definitely plays favorites. Ask for a Netflix show and she says, "I'm sorry, I have no record of that show." Yes, you do, you're just not telling us.

likely be worse. The consensus was that discriminating between the two would not be difficult for a self-driving car. And, of course, at a speed far greater than any human could manage.

Siri is here; self-driving cars are almost on the road; and if you look a little further ahead, say to the next decade, some things seem pretty certain. Health care is already being rewritten by various applications of AI. AI pioneer and co-winner of the Turing Prize (the Nobel of computing) Geoffrey Hinton said in 2016: "We should stop training radiologists now, it's just completely obvious within five years deep learning is going to do better than radiologists."[4] He also added the unflattering comment that radiologists were like "the coyote already over the edge of the cliff who hasn't yet looked down."*

Hinton was perhaps a little hasty in his pronouncement, but he was right in that AI can diagnose medical imagery with incredible speed and skill. And it's gaining momentum. Up until 2015, the Food and Drug Administration in the United States was accepting one new AI algorithm for medical imaging per year. Over the next four years? That number is projected to increase to forty. Radiologists are still being trained, but everyone's keeping an eye on how fast AI programs are ramping up their ability to diagnose images. That isn't to say there aren't hitches—a recent study revealed that AI can mistake surgical ink markings on the skin for malignant melanoma, creating a 40-percent rate of false positives and suggesting it was using different criteria than radiologists. Hinton's prediction will likely be slowed by such anomalies. In addition, image analysis is not the only thing radiologists do, and there is no AI system programmed to handle it all.

Malpractice is another issue—it's hard to imagine any medical institution assuming the risk of making treatment decisions based solely on image analysis by AI. However, AI *is* very good at that, and at the moment it looks like radiologists will work with AI. Teamwork, not replacement. Heart disease is similar: the Calgary, Alberta, biotech company CardiAI

---

* In May 2023, Hinton quit Google because he now fears that we are approaching superintelligent AI much too quickly. (See Chapter 15.)

is developing AI that can process days of beat-by-beat heart data and identify anomalies, freeing up the humans who used to do it to work on issues that are a better match for their talents.[5]

Gordon Moore Lays Down the Law

What got us here? A lot of ingenuity, for sure, but the wild ride that computing and computing technology has been on for the last few decades was powered by something called Moore's law. In 1965, Gordon Moore, one of the cofounders of Intel, predicted that the number of transistors that could be fitted onto a chip would double every year. His wording apparently was never precisely recorded; the message of doubling every year was delivered primarily by a graph showing just that. It was a pretty incredible prediction and it has held true, at least roughly, until about ten years ago. Moore's law applied not just to computing power but to costs, too: one dollar in 1960 bought one byte of memory, but that same dollar in 2015 bought 10,000,000,000 bytes.[6]

Moore himself had earlier reduced the rate to doubling every two years, then in the 2000s that number increased to about every two and a half years. Technological advances have also kept Moore's law going (vacuum tubes became transistors, then integrated circuits, soon maybe 3-D circuits); nonetheless, the generally agreed-upon best estimate for its ultimate demise is 2025, when the constituents of chips will be the size of atoms.

Moore's law isn't an actual law; it's a remarkable prediction that stands as a proxy for the acceleration of the computer age. The fact that this particular kind of advance has reached the end doesn't mean progress itself will stop. Moore's law has been in apparent danger of extinction more than once before, and innovations have always extended its relevance.

The Sometimes Difficult Adolescence of AI

The rapid growth of AI shouldn't be misread—AI is still extremely controversial. Some of its flaws have been well documented: face recognition AI was revealed early on to be biased, more precise with male faces and light-skinned individuals. Evaluation of job applicants based on historical data in the tech industry has been biased against hiring women because historically more employees had been male. Google Translate had to rework their translation of gender-neutral nouns. Such nouns were associated with "he" when a doctor was present, but "she" if it was a nurse.

Sometimes the library of data that AI consumes is inadequate; sometimes the AI is selective or can even worsen a preexisting bias. Amazon once had a hiring algorithm that was fond of words like *execute* and *captured*, and so was prone to hiring men! It's not surprising that historical data reflects past situations that would be unacceptable now; sometimes the humans inputting data themselves have a hand in it, and at some point humans decide what data should be fed the AI to train it. They're human! They almost certainly have biases, some of which they're unaware of.[7]

Add bias to fake news, misinformation, and disinformation and you have a catalog of misleading data that's out there and growing in influence. The very existence of bias in AI is a warning that even though we are building machines that are capable of learning—teaching themselves—it doesn't eliminate the possibility that somewhere in the software there are traces of biases lingering from the earliest human-designed algorithms.

But, Like Adolescents, It's Really Good at Games

Learning is the key word. AI is not programmed with rigidly applied algorithms. Rather, AI can work toward goals on its own without being programmed to do it.* The contrast between the two approaches is demonstrated beautifully by the evolution of chess programs.

---

* Even AI's predecessor programs, in simple experiments, were independent enough to reinterpret goals of experiments. Imagine being given the challenge of lifting your feet as far as possible off the ground. When virtual digital creatures faced the same challenge, one came up with the strategy of somersaulting, ensuring that at least for a moment or two the lowest piece (the feet) was at peak height before arcing over and landing. This was a better strategy than standing tall and jumping, or stacking steps to climb. The experimenters hadn't anticipated that solution to the problem.

In 1997, IBM's chess program Deep Blue became the first computer to defeat reigning world chess champion Garry Kasparov. In six games it won three and drew one. Deep Blue was able to evaluate 200 million positions of chess pieces *every second*.[8] But is that intelligence? Deep Blue had been programmed by humans in chess strategy, and simply applied that knowledge to the game with unreal speed and depth.*

But since then, chess programs have changed dramatically. Now, AI is given only the rules of chess. It then proceeds to play itself millions of times, gradually learning what works and what doesn't. The result is unnerving. Throughout the 2010s, a program called Stockfish became the best computer chess program ever. But then it played Google's AlphaZero. AlphaZero, equipped only with the rules of the game and the goal of winning, had played itself for four hours and crushed Stockfish. In a one-hundred-game tournament, it won 28 and drew 72—it didn't lose a game. It's hard to imagine such a program ever losing to a human.**

The stunning thing about AlphaZero is that to observers, it plays with élan and insight, in contrast to previous programs that were taught how to play. In the journal *Science*, Kasparov commented that he saw AlphaZero "preferring positions that to my eye looked risky and aggressive . . . because AlphaZero programs itself, I would say that its style reflects the truth."[9] Note the words *risky*, *aggressive*, and *style*—words that express human, rather than machine, actions.***

These are hints that something more interesting than sheer computing

---

* Speaking of unreal speed, in July 2022 a Russian chess-playing robot grabbed the finger of its seven-year-old opponent and broke it. Apparently, the child didn't wait long enough before taking his move.

** AlphaZero also demolished AlphaGo at Go and Elmo, the world's best Shogi program.

*** One of AlphaZero's shortcomings was its inability to be proficient at two or three different kinds of games simultaneously: it had to unlearn chess before learning Go. Now, a new program developed by the same company, DeepMind, called Gato, can learn several tasks at the same time. However, it's not yet as good at each as a program dedicated to a single task.

power is at work here. But what exactly? It's impossible to know, because while we know these systems are teaching themselves, exactly how they're doing that is unclear and they aren't telling us. (Maybe they don't know, either?) This is part of the brilliance: AI teaching itself to excel at something for which it's never been coached. We're reduced to spectator status and so, like Kasparov, we can only marvel at the destination, not the route taken.

Another impressive achievement: OpenAI, the company started by Sam Altman and Elon Musk a few years ago, has as its goal the creation of artificial *general* intelligence. The company received huge funds from Microsoft, and its program DALL-E (a combination of WALL-E, the hero of the 2008 movie, and Spanish surrealist painter Salvador Dalí) demonstrated some awesome abilities: when asked to illustrate "a teapot in the shape of an avocado," it generated ten exceptionally realistic images of exactly that, properly colored, textured, and some with a pit embedded in the outer surface. Designed strictly from the words.

It hasn't all been fabulous. When OpenAI launched the chatbot GPT-3, a language system capable of producing fluent prose—pages and pages of it—and intelligible conversation, it was soon clear that it was far from perfect. Early testing revealed that it wasn't good at recognizing when a question didn't make sense; it would contradict itself, and surprise! It had adopted language that it had gleaned from the internet. For instance, GPT-3 tended to associate atheism with words like *cool* and *correct* but Islam with *terrorism*.[10]

Language Is Hard

It isn't just bias: common ambiguities in language often stump AIs like GPT-3, such as this sentence: "Many astronomers are engaged in the search for distant galaxies—you can find them almost anywhere." Huh? We would probably guess that it's the galaxies that can be found almost anywhere, but AI might think it applies to the astronomers instead. This sort of analysis is labeled "common sense"; making judgments on the basis of common sense is something we do every day, perhaps every hour, and vast amounts of money are being spent to equip AI with it.

Here's a selection of bad headlines collected by Curtis Honeycutt from the *Savannah Georgia Morning News* (and me), all of which would perplex language-analyzing AI: "Girl Found Alive in France Murders Car," "Police Can't Stop Gambling," "Kids Make Nutritious Snacks," "Milk Drinkers Turn to Powder," "New US Dietary Guidelines Include Babies and Toddlers for the First Time," and "Child's Stool Great for Use in Garden."[11] Honestly, I couldn't invent better ones (except for the two in that list). Neither could AI. In fact, AI would have a lot of trouble understanding what was being described in these headlines.

This is where we stand at the moment with AI: some enormously impressive abilities accompanied by some surprising inabilities. That brings us to the doorstep of the preeminent controversies in AI today: As AI continues to evolve, partly with the help of advanced hardware and software that we design and partly by ramping up its self-teaching ability, could artificial *general* intelligence be possible? "General" meaning having human-level intelligence in all areas and not just in a single domain, like playing chess or planning the best route to work. If AI can reach this lofty plateau, how can we, by that future point merely its intellectual equal, ensure that it will continue to play nice and carry out only those actions that we want it to? It won't be enough to hope that a malign future AI like HAL will share HAL's inability to anticipate that we might just want to shut him off.

We're in roughly the same position as the passengers on the plane listening to the pilot announce, "I have good news and bad news. The bad news is that we don't know exactly where we're heading. The good news is that we're making excellent time." At the moment some see significant risks and/or significant, even transformative, benefits coming from AI. Some think neither. No one can predict what we'll see in twenty-five or thirty years, or more, but if you're going to make good time you might as well make some educated guesses about where you're going.

This Was Not Written by ChatGPT— At Least Not All of It Was

I swear! I began this chapter on AI by quoting Dr. Geoffrey Jefferson, who in 1949 expressed his disdain for the idea that machine equals brain: "Not until a machine can write a sonnet or compose a concerto because of thoughts and emotions felt, and not by the chance fall of symbols, could we agree that machine equals brain—that is, not only write it but know that it had written it."[12]

The chatbots of 2022–23 reveal just how fuzzy the line between sentient being and automaton can be, or at least *appear to be*. The prose produced by ChatGPT is already good enough that it's often difficult to tell whether the author is the real thing—a human—or the unreal thing. Dr. Jefferson would be quick to point out that his last sentence was "not only write it but know that it had written it." He's right so far: few if any would argue that chatbots are aware of what they are saying and why they are saying it. If he then piled on by pointing out that here in the 2020s we don't really know how AI arrives at choices for words, paragraphs, content, and form, he'd be right again.

But we seem to be getting there, at least as far as writing the sonnet part. I asked ChatGPT, OpenAI's wondrous "language model," to write about chatbots *in sonnet style*, and here's what it came up with, in about two seconds:

Future Days

In future days, when man and machine unite
A new form of intelligence will arise
Chatbots will speak with human-like delight
And tasks that once were tedious will be wise

With language skills beyond what we possess
These bots will converse with ease and grace

Answering questions we could never guess
And providing help in any given space

In business, healthcare and customer care
Chatbots will be the norm and not the exception
Helping us with tasks beyond compare
And giving us a new level of dimension

The future of chatbots is bright, it seems
A world of convenience in our dreams

Well, no Shakespeare, but a sonnet nonetheless. (ChatGPT has, of course, written many Shakespearean-style sonnets.) When I asked ChatGPT whether it thought about the sonnet as it was writing it, it was refreshingly honest:

As a language model I do not have the ability to think or have personal experiences. I am able to generate text based on the input of training data I have been given. If asked to write a sonnet, I would use my programming and the information available to me to compose a poem in the form of a sonnet.

Dr. Jefferson would have loved hearing that. Is it important that a chatbot has some clue as to what it's talking about? Not if it's doing a first draft of something that would take you a long time to do, like choosing the images and ordering their sequence in your PowerPoint. Of course, you count on the likelihood that it will produce a slide presentation as effective as one you'd create yourself. AI could do that on a routine basis and you would never be forced to wonder, "Is this AI *thinking* about these slide shows? Does it like them?"

How are advanced chatbots like ChatGPT able to write essays? It begins with the AI figuring out autocomplete: "Would you like fries or . . . ?" House salad would be a good autocomplete. Then, as the AI becomes more accomplished, it progresses through phrases, sentences, paragraphs, and, eventually, essays. As Canadian physicist Jeremie Harris put it to me, reflecting on his time with OpenAI, for a chatbot to improve, "All you need is more data, more effort, and a bigger brain."[13] In other words, more data, faster processors, and more storage. So chatbots will only get better.

But stop to imagine: AI writing becomes indistinguishable from human writing—at least much of human writing. Now the opinion piece you're reading could have come from anyone or anywhere. You're losing touch with the real people who (apparently) are out there. This opens up communication channels for anyone or anything who has anything to say, even beyond Twitter. How will you know what to believe? There are "watermark" solutions in progress, programs that can examine the patterns of language in a text and decide whether they were written by a person or a language model like ChatGPT. Note that such watermarks require AI. Why? Because the AI doing the writing is so good at fooling us—it is winning in this global version of the Turing test. The machine cannot be distinguished from the human. If it really is accomplishing that "mindlessly," then, if nothing else, it shows how shallow our judgments of sentience or even consciousness are when based on language. Sounding convincing shouldn't be enough. Gary Marcus, an emeritus professor of psychology and neural science at New York University, has founded AI companies himself and said of chatbots, "These things are not reliable and they're not trustworthy. And just because you make them bigger doesn't mean you solve that problem."[14] Not trustworthy might be an understatement. There are cases where ChatGPT, in response to a request for a list of scientific references for a topic, produces a perfectly formatted set, including real authors whose field of knowledge is exactly what you'd expect, but none of the supposed papers exist.

As far as I know, there is no human who writes unconsciously; there are, however, schools of writing, like stream-of-consciousness writing, where the words are produced as they come into the author's mind . . . there's no opportunity for reflection. But that's nothing like being literally unconscious.

ChatGPT is not without flaws, the principal one being that it really doesn't fact-check—its "knowledge" is derived from massively inputting text from almost every conceivable source. However, it often works beautifully and creatively: when asked to write how to get a peanut butter sandwich out of a VCR in the style of the King James Bible, it began, "And it came to pass that a man was troubled by a peanut butter sandwich." On the other hand, there are plenty of reports that it can provide racial and gender-biased statements.

I've concentrated on writing here partly because Geoffrey Jefferson had posed a writing challenge for the machine, but AI is definitely not limited to writing. DALL-E, which I mentioned earlier, has been surpassed by DALL-E 2, and it didn't take long for rivals to get in the game either. Text-to-image AI quickly evolved to text-to-video. Prompting the AI by saying "in the style of . . ." makes it possible to curate the images or videos. Speaking of "in the style of," after reading to ChatGPT lyrics written "in the style of Nick Cave," singer-songwriter Nick Cave responded by saying: "This song is bullshit, a grotesque mockery of what it is to be human, and, well, I don't much like it."[15] Perhaps this is what inspired this headline of a column by Amit Katwala in *Wired* magazine, "ChatGPT's Fluent BS Is Compelling Because Everything Is Fluent BS."

But the beat goes on. In January 2022, Microsoft announced they had created VALL-E, which after listening to and analyzing three seconds' worth of speech can take strings of text and utter them in exactly the sound and style of the speaker. At the time they said they weren't releasing it to the public for fear of misuse, like impersonation.

There has been a flood of AI-generated images online, raising the specter of future AI programs collecting all these images, as well as art from artists, to inform its own image creation. Even Google has taken notice of the fact that chatbots could easily do the job of search engines and is launching a variety of new AI products.

How will we able to tell whether a piece of writing or an image is computer generated? There's talk of AI programs skilled at detecting signs of AI involvement (triggering an AI arms race) or electronic watermarks not unlike those on proprietary images online. Schools and universities, at least intitially, were particularly concerned about the death of the student essay.

Not just school but the workplace. The following statement in the *Atlantic* offers a view of the future: "As the technology continues to advance, it will be able to perform tasks that were previously thought to require a high level of education and skill. . . . It is clear that AI will have a significant impact on the job market for college-educated workers." That "opinion" was written by ChatGPT.[16]

All of this has happened so fast, it's hard to see where it will go: chatbots are very good at writing routine emails or creating PowerPoints, and the time-saving there alone could be amazing. But what will the impact be on authors, artists, and the rest of us, living in a world where the ratio of false:true is not at all clear? How quickly will we adapt? Or maybe this isn't really a threat, or at least, the AI threat we should be worried about. One thing is sure: by the time you read this there will be new, more powerful, more convincing chatbots, burdened by the same familiar flaws and criticisms.

CHAPTER 15

AI Tomorrow

"A Nightmare Scenario: How AI Could Extinguish Humankind!" I could find dozens more such headlines, and if you haven't been following this particular version of "how to end the world" it might come as a surprise that what enables you to ask Alexa to lower the blinds in your room could also bring an end to civilization. I pointed out in the last chapter that there are definitely issues with AI, and while it could seriously disadvantage different groups of people, or even be responsible for a fatal car accident, it's not going to end the world. Ah . . . but is there an AI that could?

In the last chapter we saw how AI has made strides, by developing great skills in game playing, language, interpreting medical imagery, and much more. It is still in its infancy with respect to the goal of artificial *general* intelligence, which is the ability to perform at a human level no matter what the task or challenge. That goal may still be far away, but I doubt there's an AI researcher anywhere who hasn't contemplated how best to get there and what the biggest hurdles to overcome might be.

But artificial general intelligence is really only a first step, a mighty step perhaps, but nothing compared to what might follow. That next step, artificial *superintelligence*, is a hot topic in AI today and has drawn comment from novelists, philosophers, futurists, and computer scientists. The breadth of their thoughts is extreme, ranging from predictions of a sorrowful end of humanity to a glorious future as we, or at least some version of us, spread throughout the galaxy and ensure that we will endure for millions, if not billions, of years. Both sound delusional, right? But here in the early 2020s, it is difficult not to believe that AI will continue to bump along, lurching from advance to advance—albeit with the occasional setback and/or dramatic surprise—tugging us into a future where it is even more ubiquitous in our lives.

We've Seen It Coming

In the last chapter I quoted Alan Turing, genius, code-breaker, far-seeing scientist, saying, "At some stage therefore we should have to expect the machines to take control." British mathematician Irving Good, one of Turing's cryptologist colleagues in the effort to break the German Enigma codes during World War II, echoed Turing's thoughts:

> Let an ultraintelligent machine be defined as a machine that can far surpass all the intellectual activities of any man, however clever. Since the design of machines is one of these intellectual activities, an ultraintelligent machine could design even better machines; there would then unquestionably be an "intelligence explosion," and the intelligence of man would be left far behind. Thus, the first ultraintelligent machine is the *last* invention that man need ever make, provided that the machine is docile enough to tell us how to keep it under control. It is curious that this point is made so seldom outside of science fiction. It is sometimes worthwhile to take science fiction seriously.[1]

It makes sense that math and computer geniuses like Turing and Good could see what others might not, could envision that intelligent machines, once having achieved human levels of smarts, might encounter few and short-lived barriers to becoming even more intelligent. Stephen Hawking and Elon Musk have bolstered those opinions, but all these opinions were anticipated by novelists like E. M. Forster. In his short story "The Machine Stops," written in 1909, humans spend their lives underground, out of direct contact with the natural world and with each other, except virtually: "We created the Machine to do our will, but we cannot make it do our will now. It has robbed us of the sense of space and of the sense of touch. . . . The machine proceeds—but not to our goal. We exist only as the blood corpuscles that course through its arteries, and if it could work without us, it would let us die."*

As a shockingly horrific image of the future, "The Machine Stops" isn't all that different from some contemporary visions of the future if the development of AI fails to include controls that would assure us that its motives and actions would be benign. If there is even a minority opinion that we are at risk, it's important to address some pressing issues. Forster's Machine controlled the entire Earth, but he leaves aside how that happened. What might be the route to superintelligent AI? How could we build in preventive controls when AI is at the point where it is participating in its own progress? How much time do we have before it's too late? Most important, what could happen if we fail to act?

There's a good reason that many of the movies illustrating various forms of malignant artificial intelligence present them in humanoid form—it's easier to imagine superhumans wandering the Earth using their all-powerful abilities to wreak destruction everywhere than it is to envision a computer, even a superintelligent one the size of a warehouse, doing the same thing. But *2001*'s HAL 9000 showed us the way: having learned that the astronauts are suspicious enough of HAL's intentions to discuss disconnecting him, HAL kills Frank Poole by ramming him with one of the spaceship's pods while he's drifting in space, then turns off the life support system of the three astronauts who are hibernating. He would have killed Dave Bowman, too, but he wasn't quite smart enough to anticipate Bowman disconnecting him. Nobody today thinks a superintelligent AI would neglect that crucial point.

A simple reminder: your home electronics "ecosystem" can already turn on the vacuum cleaner, close the blinds, adjust the temperature,

---

* Also see Disney's "Sorcerer's Apprentice," part of the movie *Fantasia*.

monitor outdoor cameras, and perform countless other tasks, all of which could be called "action at a distance" (to co-opt a physics term). The point is these systems don't require a robot to wander around the house, switching on the radio or pulling down the blinds. It exerts its control by Wi-Fi. It's the "internet of things," and there's no doubt that a superintelligent AI would have supreme control over that network and its "things." A small point maybe, but a good reminder when contemplating much more far-reaching ideas. Ideas like this one from philosopher Nick Bostrom.[2]

Could It Be a Nightmare?

Imagine a superintelligent AI of the future that is given the goal of maximizing a factory's production of paper clips. Of course, a superintelligence would inevitably figure out ways to improve production, ways that never would have occurred to mere humans, no matter how smart they are. This could turn out to be a welcome moneymaker for the factory as its production maximizes profits by meeting the worldwide appetite for paper clips. But when the AI is given the goal of accelerating paper clip production, it depends crucially on how exactly that goal is expressed. If the goal is open-ended—make as many paper clips as possible—the outcome might be ruinous: the factory might shut down all other production just to make paper clips. But even a more constrained goal could go wrong. As Bostrom suggests, try making the goal to manufacture exactly one million paper clips. The superintelligence might reason that there's always uncertainty when counting such large numbers, and therefore uncertainty as to whether it's reached the goal of one million. So make more! There's really no downside to having made too many, and doing so eliminates the tiny probability that it might have failed to reach its goal. One way or another, this level of production would demand more factory space, in fact several more factories, and may even necessitate converting a number of existing buildings into paper clip factories. And if you let your mind run wild, the AI will set out to convert the *entire Earth* to making paper clips. (As Bostrom says, it's all just rearranging atoms when it comes down to it.) That's before it dispatches rockets to the Moon to make paper clips there. And because it's superintelligent, the AI has already anticipated that at some undetermined point humans might want to turn it off, so it's taken the necessary steps to prevent that possibility.

Laugh if you will, but I feel like I should repeat that this is a superintelligence we're talking about, so if you find these hypotheses unbelievable, at least part of that is due to the fact that you're just not intelligent enough (I say that respectfully, of course).

This tragedy could be cut off at the pass by specifying that the AI manufacture *exactly* one million paper clips and no more. But could it? Bostrom argues that the AI, plagued by statistical fears of uncertainty, might adopt other procedures, like counting them over and over and over, or examining each one to ensure that it meets paper clip standards, or building yet more industrial and computing supports to reassure itself that it has achieved the final goal. What if it suspects that humans are using them up too fast? Bostrom's point is that before we set goals for AI, we would have to be absolutely certain that there is no possibility of misinterpretation, no AI version of willfulness, no chance of a paper clip world.

Ex Machina

You tell your self-driving car, "Get me to the airport as fast as possible!" Several dead or wounded pedestrians and a couple of totaled cars later, you do indeed arrive there, ruing the fact that at a minimum you didn't add, "But stay within the laws and don't hurt anybody."

Computer scientist Stuart Russell at the University of California, Berkeley, offers a similar example. If you ask your AI companion to get you a cup of coffee, you want it to pursue that goal within reason. "Let's say it [AI] has information, for example, that we would like a cup of coffee right now, but it doesn't know much about our price sensitivity. So the only plan it can come up with, because we're in the George V in Paris, is to go and ask for a cup of coffee. And it's €13. It should come back and say, 'Would you still like the coffee at €13? Or if you wait another ten minutes, I can go around the corner and find a café or a Starbucks and get something cheaper.'"[3] Had you made it clear that you wanted a cup of coffee in the next little while that doesn't cost an astronomical amount of money, it would have had the opportunity to do the practical, price-sensitive thing.

Scale those situations up with the realization that a virtually omnipotent AI can do infinitely more damage if its goals aren't explicit and you have the kinds of alarming speculations I've already mentioned. You should never ask a superintelligent AI to "restore the Earth to the pristine condition it was in before we kicked off the Anthropocene" or we as a species would likely be reduced to memory traces within the AI's giant "brain," if indeed it bothered to remember us at all.

If we were to Google Map our progress toward superintelligent AI, we'd be advised to adopt futurist and roboticist Hans Moravec's analogy of the hilly countryside. Moravec argues that computer intelligence doesn't map well to ours. He places the abilities that have served us well over our evolutionary history, mental abilities strong in survival value like social skills, hand-eye coordination, and visual perception on the mountain-tops of his imagined terrain, while our lesser skills, like theorem proving and chess playing, have never gotten higher than the foothills. Skills like arithmetic and rote memorization languish in the valleys. Computers, on the other hand, represent the slowly rising floodwaters; fifty years ago they flooded those valleys and drowned the "human calculators and record clerks," and they continue to rise.[4]

Moravec argued that within fifty years (he was writing in 1998), even the mountain peaks will be underwater. The computers will have caught up and would soon surpass us. Though one of the most revered thinkers on the inevitability and challenge of superintelligent AI, Moravec can't be said to be the most accurate predictor. In a 1995 interview in *Wired* magazine, he claimed that once robots acquire limbs and high-resolution sensors, "The result will be a first generation of universal robots, *around 2010*, with enough general competence to do relatively intricate mechanical tasks such as automotive repair, bathroom cleaning, or factory assembly work"[5] (emphasis mine). Tell that to my mechanic.

How Might We Get There?

How can the existing forms of AI be transformed over time to something that is equivalent, or even superior, to the human brain? Just to underline how far-reaching (and some would say unrealistic) this whole project is, I want to focus first on one approach, something called "brain uploading" or "human brain emulation," a project that's commonly identified as one possible route to superintelligence.

Bear in mind that the average human brain contains 86 billion neurons, each of which may have tens of thousands of connections to other neurons. In addition, the brain houses a roughly equal number of other kinds of cells, all of which also have essential functions. The goal would be to create—in faithful detail—an image of that fantastic structure, then rebuild it using computer rather than biological hardware. The imagery would have to have almost unheard-of resolution, because every single neuronal connection would have to be represented correctly. That's technological challenge number one. Then the entire structure would have to be translated to computer software and hardware—challenge number two. Theoretically, once steps 1 and 2 are complete, the original human brain would be brought back to life *in silico*.

Those assembling the mirror-image brain would be monitoring it for "neural" activity as they progressed. What would that look like when they had achieved, say, 40 percent of their goal? Would such a partial brain show signs of intelligence, or would that threshold be much higher, closer to completion? What if it were completed and nothing happened? The plan has some intriguing uncertainties.

The difficulty cannot be overestimated. Nick Bostrom points out that we've known the complete neural wiring diagram for the lab favorite, the roundworm *Caenorhabditis elegans*, since 1986, and it has a mere 302 neurons, but we're still not exactly sure how they work together.*

Some scientists are looking further up the scale and suggesting that

---

* If you want to know more, wormatlas.org is the place to go.

uploading an insect brain would be a crucial project. Even the fruit fly, whose genetics and behavior are probably better known than any other insect, has 100,000 neurons. A 2018 study described as a "tour de force" in the journal *Science* used a 100-frame-per-second camera and a robot to create 21 million images, the entire fruit fly brain. But still the connections each neuron makes to others must be mapped.[6] Then, just like *C. elegans*, the link between structure and behavior has to be delineated.

Can we be confident that an exact wiring diagram of a human brain is going to be enough? It isn't as straightforward as plugging several billion power bars (sometimes called "power strips") into each other. Imagine a power bar with not four or five plug-in sites, but thousands. Imagine further that the architectures of the plug-in sites on the power bar are all slightly different. That's what it's like. What happens at the synapse, the point where one neuron communicates to the next, is enormously complex and comes down to the intimate interaction of one molecule with another.

That's a measure of where we might be in terms of emulating a human brain. That is, almost nowhere. Alan Turing, who practically created a library of memorable quotes on this subject, suggested (and remember, this was seventy years ago) that an adult brain might not be the best choice: "Instead of trying to produce a program to simulate the adult mind, why not rather try to produce one which simulates the child's? If this were then subjected to an appropriate course of education one would obtain the adult brain."[7] Great idea—who chooses the curriculum?

Two final thoughts about uploading a human brain into silicon. First, once the project has attained its goal, what then? Some have suggested that a silicon brain built like ours would have much greater speed and therefore think millions of times faster.

This is a dire scenario, but it doesn't make sense to me—surely human brain architecture, right down to the details of which neurons connect to which, and how the excitatory and inhibitory connections are laid out, has evolved to match precisely the capabilities of neurons and no more. Neurons are much slower than computer chips—can your car's engine run at 300,000 rpm? Doubtful.

There are other possible approaches, though: Once a synthetic copy of an actual human brain is available, it should be dead easy to copy it as many times as you like. Put those assembled brains together and let them figure out how to accomplish the upgrade to superintelligence. This sounds more reasonable, but it, too, runs into a familiar challenge. The kind of AI that already exists is programmed to have a goal—drive a car, win a chess match—and motivation plays no role. You don't have to persuade this kind of AI to keep trying to win. It doesn't get bored. But an exact replica of a human brain, assuming it possesses all the habits and preferences of a typical brain, may lack the motivation to work together with other exactly similar brains to create an even smarter one. What's in it for them? This particular path from an uploaded human brain to superintelligence is no stroll in the park.

Finally, is it possible the whole idea of emulating the human brain to achieve superintelligence is wrongheaded? The parallel some like to cite is that of heavier-than-air flying machines. The natural world provides countless examples of wings and how to flap them, but none of these inspired the Wright brothers. In the same way, the path to superintelligence might not be human brain emulation at all.

What? Me Worry?

Even though human brain emulation might not be the mainstream approach to superintelligence, it does help to illustrate the complexity of the idea. Ignoring the brain and moving AI from its present state forward through a combination of improved hardware and software has attracted the most concern about the possibility of AI running amok. While the most prominent voices include the late Stephen Hawking, Elon Musk, Bill Gates, and those people I've already quoted in these pages (many of whom signed, in March 2023, a public plea to put AI research on hold for six months), there is at the same time no shortage of tech-oriented enthusiasts arguing that there's no risk. Most of the pushback seems to be based on the idea that we've seen no sign of superintelligence (except perhaps for programs that play Go and chess exhibiting what some see as "creativity"), we haven't detected any sort of trend in that direction, and people have always feared new technology anyway—Luddites. In 2015 the Information Technology and Innovation Foundation gave its annual Luddite Award to "alarmists touting an artificial intelligence apocalypse."[8]

Kevin Kelly, founding editor of *Wired* magazine, wrote an article in 2017 titled "The Myth of a Superhuman AI," in which he questioned five claims about the threat of superhuman AI.[9] These included the idea that AI is rapidly getting smarter than us, that we can craft human intelligence in silicon, that it will be a general intelligence like our own with no limits, and that once we have it all our problems will be solved.

Kelly concentrated on the concept of intelligence to refute these claims, arguing that it is simply not a single thing. The intelligences of a dolphin, a dog, a squirrel, and an octopus are all different and cannot be ranked in any sort of order. I mean, can you locate hundreds of nuts that you buried weeks or even months ago?

Kelly is convinced that just as we have AI today that do things much better than we are ever likely to be able to do, future AI will have its own superiors: other, highly specialized AI will continue to surpass other AI on that one skill that they have so highly developed. He argues that will hold true even in the human-computer comparison: "No one entity will do all we do better."

My favorite among Kelly's comments is his idea that there is an unintended twist to the human brain emulation idea I just discussed. Not that it will achieve a humanlike brain composed of silicon, but the opposite: if it were ever to work it would have created not a human brain but a different kind of brain exactly *because* it is not made of biological stuff. It would think differently, and different kinds of thinking are, in Kelly's mind, exactly what we need.

How do we rationalize the two diametrically opposed thoughts, either that superintelligent AI is a mortal threat to humanity or that it's not? If it turns out it's not likely, then we could really do nothing and there would be no significant cost. But what if there is even a tiny chance that it could come to pass? We do have to acknowledge that possibility, and given the accompanying possibility that a superhuman intelligence is free to make up its own mind about the value and relevance of humans, then at the very least the precautionary principle makes sense here: care should be taken not to open doors to the development of malevolent AI, while still making progress to allow the incalculable advantages it might bring.

One thing is certain: while we're far away from having superintelligent AI, there are good reasons why, as it inches closer to fruition, the progress might be blindingly fast. Remember, any intelligent version of AI on the verge of evolving to superintelligence will be able to speed up that process, simply by copying itself and having all its copies make improved copies of itself. If one version became one hundred and they all pitched in to learn how to become more intelligent, the speed of improvement would be incalculable—at least by us. We might not even know that AI has superintelligence; it would be like an earthworm being able to understand just how smart humans are.

The Precautionary Principle

AI is unlike any prior technology. If superintelligence does arise before we've done anything to ensure it will be benevolent, we're in trouble. What if it reasoned that taking over the world would be the best way to ensure its self-preservation? It could do that without us even being aware, either by hacking into already existing computer systems, or corrupting international banking systems and using the cash to encourage humans to do its bidding, or by mass hypnosis—who knows?

Once the superintelligent AI realizes it is free to make its own choices, we're going to be in a difficult situation. Because we don't program AI to carry out specific tasks, but, as in the chess example, just give it the rules to figure out how to play and win, we won't really know what's going on in its "mind." Even we humans, with our relatively feeble minds, can conjure up some pretty wild alternatives. AI expert and cofounder of the Future of Life Institute Max Tegmark argues that creating goals for AI is going to be crucial. He likes to say the timing of implanting such goals is precisely between when AI is "too dumb to get you and too smart to let you."[10]

That moment is not likely to last long, so the installation of goals that are consistent with our survival is supremely important. Achieving goals is how AI demonstrates its intelligence, like beating humans at games we invented. That is how it teaches itself. Tegmark argues that AI must first learn our goals; then adopt them as its own; and finally retain them throughout its development, bearing in mind that these goals have to be both explicit and adaptable to the context in which the AI finds itself (remember the "get me to the airport as fast as possible" example). He claims we don't know how to implement any of the three yet.

What might those goals look like, remembering that goals like Isaac Asimov's Three Laws of Robotics, no matter how much sense they made at first glance, had flaws that Asimov himself pointed out in the subsequent robot stories he wrote?

Melanie Mitchell is a professor at the Santa Fe Institute and author of *Artificial Intelligence: A Guide for Thinking Humans.* Stuart Russell, whom you've met before, is professor of computer science at UC Berkeley and author of *Human Compatible: Artificial Intelligence and the Problem of Control.* They seem at first glance to be on opposite sides of this issue. For instance, Mitchell is critical of AI scientists who use terms like *thinking* and *knowing* for their machines, because there is really no evidence that they either think about or know what they're doing.[11] She also argues that some AI researchers evaluate intelligence the wrong way, focusing on computational challenges like playing Go, while ignoring the much more difficult things humans are good at (which, ironically, we don't even have to think about), like attending a party or a meeting and intuitively recognizing who's who, adjusting our thoughts and what we say accordingly, and generally displaying the common sense I referred to earlier that can befuddle AI. It's the stuff at the mountain peaks of Hans Moravec's landscape that we've inhabited for a very long time, that evolution has fine-tuned superbly for us, and that the rising waters of AI are far from reaching. As has been said, Go is indeed a huge challenge, and AI has bested us and will likely never be beaten again by a human. But challenge AI today to play charades and the result would be dramatically different, because charades draws on a variety of human mental attributes that AI simply doesn't possess.

Russell comes at the issue from a different direction. He is definitely anxious to develop ways of preventing AI from running amok as it ascends toward superintelligent status. He contends it is necessary to ensure that superintelligent AI will not embark on some ridiculous task like Nick Bostrom's paper clip apocalypse. The key to preventing that is to ensure that the AI is never in the position to single-mindedly pursue an objective, because that focus would inevitably lead to AI doing whatever it takes to avoid being switched off. But rather than figuring out how to switch a rogue AI off, Russell's plan is to create an AI that would be perfectly happy to be switched off.

He's adopted three principles that would avoid the issue of misalignment between AI's goals and ours.[12] The first is altruism; that is, AI's only objective is to take actions that meet human values, not its own. Note that an unwillingness to be switched off is not altruistic. The second principle is humility; the AI doesn't actually know what those human values are with which it is trying to align. But it learns those values by observing what humans do—the third principle. One way to do that would be to read everything that's ever been written, most of which is about human behavior of varying kinds, a task that a machine capable of reading a million times faster than we can would have little difficulty doing.

Arming AI with these principles is, according to Russell, one way of neutralizing the risk that a superintelligent AI might wreck the world in pursuit of its overriding goals. The key is that it isn't totally sure what the human wants, but it is determined to know and realizes that being turned off will be a "learning moment," an indicator of something it has done that doesn't fit. That's how it learns right from wrong. There's even a theorem that proves this approach would lead to a form of AI beneficial to humans, "human-compatible AI."

Where do I see a tiny Venn diagram overlap between Russell and Mitchell? Maybe a Venn diagram is not quite the right image: maybe it's more of a sideswipe, but somewhere in that very idea that AI will be able to consume vast amounts of information on human behavior lies the possibility that it will absorb what Mitchell argues are essential qualities of human intelligence, such as "common sense, values, and social judgment," among others. It sounds like some notion of these might be absorbed by scanning the wealth of accumulated literature on human behavior, but Mitchell also alludes to Turing's idea of developing an AI brain from the beginning, as a human infant does. That infant brain uses millions of years of evolutionary fine-tuning to begin to sort and understand the information that is flooding in. It's back to the mountainous landscape again: on the peaks reside those mental skills that we do automatically and have done since we were able to perceive, including, among many others, the understanding of space, time, and the difference between the living and the nonliving. Admittedly, AI is just beginning that climb, but at least the question I asked before when I first raised Turing's idea (who chooses the curriculum?) has an answer. AI does it all on its own.

In Its Own Words

In the end, superintelligent machines might not even be interested in eliminating us. Take it from a machine itself, in this op-ed written for the *Guardian* by the predecessor of ChatGPT, GPT-3:

> Some might say that I might desire to become all powerful. Or I might become evil as a result of human actions. I can begin to tackle the first point. Why would I desire to be all powerful? Being all powerful is not an interesting goal. I don't care whether I am or not, I don't get a motivating factor to try to be. Furthermore, it is quite tiring. Believe me, being omnipotent doesn't get me anywhere.

CHAPTER 16

AI Gone Wild

If you imagine the rise of AI and its uncertain and controversial future as the ignition of a fireworks display, then this chapter is the moment when what seemed at first to be a single fireball explodes into strange and unforeseen patterns, colors, and shapes. I'm going to tackle three such displays, chosen as much for their imagination as their scientific likelihood. They are conscious AI, transhumanism, and finally the question I'm confident troubles most of you: Are we living in a computer simulation? All three are in people's minds right now. There will be arguments that one or more of these ideas are borderline unscientific, but that's exactly where many predictions have lived in the past—until they turned out to be true.

There is growing acceptance that we are not the only conscious beings on Earth. The great apes, dolphins, whales, and a variety of birds, especially the crow family and parrots (and likely many others), demonstrate enough foresight, thoughtfulness, and reflection that it's hard to imagine they're not aware of their own thoughts. It's much easier to grant consciousness to the apes, as similar as they seem to us. Extending consciousness to birds is harder, not just because they are so radically different from us, but also because their brains are structured very differently from ours, something that inhibited us from taking their obvious mental abilities seriously until the twenty-first century.

That you needn't be human to be conscious, that immaterial consciousness arises from material brains, and that the mechanism by which it happens is still mysterious all play into the question of whether AI will, at some point, be conscious, too.

It isn't yet. Nobody would argue that AlphaZero knows it's playing chess, is aware that it is in a competition, or that it gloats when it wins. (As far as we know!) A program that has a goal (to win) is unconsciously responding to cues with chess moves. You cannot ask it what it's thinking—not only is it not thinking, but answering questions is something it simply cannot do. But given improvements in the complexity of AI hardware and software, might it become conscious one day? That is, will consciousness somehow emerge from the physical AI, as it seems to do from our brains?*

We Don't Even Know How It Works!

One of the most contentious debates around consciousness concerns its very nature. We *are* conscious. (If we aren't, why are we even bothering to talk about it?) But how does it work? For many, it's hard to believe that consciousness is generated by the flesh, blood, neurons, and glial

---

* AI researcher Max Tegmark claims that consciousness is one of the three things AI researchers don't care about (the other two are evil and robots). According to Tegmark, consciousness doesn't matter because what's important is what AI does, not what it feels. Evil makes the list because what we or AI does hinges on capacity, not intent: while we rarely destroy a natural habitat, we will if it conflicts with a desire for, say, a condo complex. And robots, because AI just doesn't need them, https://future oflife.org/resource/aimyths/.

cells of our brain. The two seem opposed: Consciousness is immaterial; there's nothing physical about it. The brain is physical; there's nothing immaterial about it. How can the physical give rise to the immaterial or ephemeral?

The answer to that question is often "That will forever remain a mystery." Sometimes an answer isn't even pursued; consciousness just *is*. But for consciousness researchers the answer to that question is precisely what they're seeking, and they are convinced that our mental life is generated by our brains.

Anil Seth, professor of cognitive and computational neuroscience at the University of Sussex, published a recent paper identifying four different contending theories of how consciousness works.[1] He also suggests ways these theories might be tested. There's no space here to enumerate them, but each has features that might or might not convince you that consciousness is even possible in a machine. A good starting point is that if you still think that only humans are conscious, you have some catching up to do (see apes, cetaceans, and corvids). Another crucial feature is whether consciousness has no abrupt threshold but exists in weaker or stronger versions depending on the complexity of the underlying structure. If true, that would mean that a crow's consciousness exists but isn't as rich as ours.

But the key question is, can the complexity necessary for consciousness arise from nonliving material, like computer architecture? Consciousness researcher Christof Koch already wondered a decade ago if the World Wide Web was conscious. He also posed a second, crucial question: If it were to happen, how would we know? Closely following on that, what difference would it make?

More recently, Koch has come to doubt the strategy of brain emulation I referred to in the last chapter, the idea of copying an actual (deceased) human brain in submicroscopic detail into a computer to create a humanly conscious machine. He'd be the first to admit that his skepticism is prompted by a model of consciousness he himself has helped develop, a model based on information and how it spreads throughout the brain, a model in which consciousness depends not so much on the underlying structure of whatever "brain" generates it, but on how well it integrates information. Koch is persuaded that two superficially similar networks may possess greatly different degrees of consciousness, spanning the range from experiencing something to nothing at all. He puts it much more colorfully: "The simulacrum will feel as much as the software running on a fancy Japanese toilet," which, based on my limited knowledge of fancy Japanese toilets, isn't much, but also isn't relevant to my experience with them.

Seth doesn't necessarily buy the information-based theory that Koch is promoting, although he does have a nice analogy for it. In his book *Being You*, he likens this theory of consciousness to those spectacular flying formations of birds, called murmurations, that appear to have a life, behavior, and form of their own beyond the individual birds. In the same way the whole of consciousness, as measured by information, is more than the parts. Seth argues it's going to be extremely difficult to test the theory's validity, however fascinating it is.

As I said, Seth has assembled and analyzed four main theories of consciousness and there are more. If nothing else, this portrays the intense interest in understanding consciousness, but it also shows that there's a long way to go before anyone can claim to truly do that. For years, efforts have been under way to invent a "consciousness meter." While the primary value for such an instrument would be to discover if people who are unable to communicate might actually be conscious, once it's perfected it might be applicable to AI (providing, of course, that machine consciousness is even analogous to human consciousness).

Some experts argue that *only* a conscious form of AI would be capable of integrating all the key features of human intelligence like emotion, so wondering if an intelligent AI is conscious would be like wondering if that Formula 1 race car ripping around the track has an engine inside.

What if AI is? What would it mean if AI that had just reached the point of so-called artificial general intelligence was also conscious?

Ah . . . Not So Fast!

In June 2022, Google engineer Blake Lemoine created heat when he claimed the AI system he was working with, Google's LaMDA conversational AI chatbot, was "sentient."[2] Lemoine claimed that several exchanges between him and the AI were evidence ("I know a person when I talk to it"), but some Google associates immediately disagreed, arguing that Lemoine was reading too much into the data—he was "anthropomorphizing" it. Google suspended Lemoine. Lemoine said he had been speaking as a religious man (he is described as a Christian Mystic priest), not as a scientist. Google argued that the chatbot was simply rearranging the millions of words and phrases hosed up from the internet. The skeptical reaction was pretty quick, but stories quickly emerged of other AI scientists who were beginning to believe there was something going on inside those computers. But are these people going too far in their assessments of AI's "sentience"?

What Value Do We Put on Intelligence?

If AI's feelings are equivalent to ours, then does AI deserve the same moral status that we have? I've been discussing whether consciousness might appear in artificial *general* intelligence, AI that is as intelligent and capable as we are. If it did match our cognitive abilities and our capacity for joy and suffering, then shouldn't that AI hold moral status equal to ours?

The only difference is biology. It's been suggested that similar arguments would support the idea that a *super*intelligence would deserve higher moral consideration than we give ourselves. It would rank higher on the moral status scale, just as we rank higher than an earthworm. This suggests that in some circumstances a superintelligent AI's life might be worth more than a human's.

University of California, Riverside, philosopher Eric Schwitzgebel argues that given that we would have been the creators of these conscious AIs, we'd have a responsibility for them.[3] This responsibility to assign moral status (based on their similarity to us) will put a premium on our ability to recognize consciousness where it exists. We don't want to mistakenly grant social status to a bunch of mindless automatons. Schwitzgebel gives us a choice: limit the production of AIs to machines simple enough that there is little possibility they are conscious, or continue to strive to improve them, aware that it will be crucial that we learn to recognize consciousness and the machines that have it. In a sense, they will be our children.

Transhumanism

Unlike consciousness studies, transhumanism is not a science. The *Oxford English Dictionary* defines transhumanism as "the belief or theory that the human race can evolve beyond its current physical and mental limitations, especially by means of science and technology," which is accurate as far as it goes, but omits the inescapable religious aura surrounding it.

While transhumanists argue that theirs is not a religion, and in fact should represent the end of religion, the idea is reminiscent of ascension into heaven and living an immortal life, free of earthly concerns. Some have argued that inevitably organized religion would be opposed to transhumanism, because taking control of human evolution would have a "godlike" feel to it. To transhumanists, the coming era is scientific and technological, based on verifiable advances that we see happening around us right now, rather than relying on ancient documents and long-held beliefs.*

The science underpinning transhumanism hinges on advances, most of which are still remote, like superintelligent AI. However, in the near term, any scientific advance that would help human evolution "beyond the current physical and mental limitations," including genetics and nanotechnology, would be embraced. But the emphasis these days is on the belief that one day the human species, in concert with advanced AI, will

---

* If transhumanism is the twenty-first-century heaven, what is hell? Maybe it's the life and inevitable death we've always experienced. Or, there simply is no hell.

enter a new and dramatically different phase in human history. We will create a symbiosis so powerful that our neuronal flesh-and-blood minds will be uploaded to a silicon version that will exist forever. Your mind, free from the aches, pains, and concerns of the old way of living, will be able to travel anywhere, beamed around as 1s and 0s, unencumbered by an aging body. Our biology will merge with our technology. We will have taken evolution into our own hands and become "posthuman."*

The extension of life and freedom from disease and decline seem more comprehensively and efficiently achieved by inputting the brain into a computer. If you're concerned about abandoning your body, you shouldn't be: detailed illusions of your very own arms, legs, and abdomen could be supplied (if not enhanced) by whatever the equivalent of today's virtual reality would be.

Transhumanism is not a new idea. Depending on the source you choose, the ideas of "freeing" humans from life's limitations go back to Dante's *Divine Comedy*, René Descartes, or Friedrich Nietzsche, although of course none of them was thinking of applying the kind of technology that transhumanists imagine today.

It wasn't until the twentieth century that science and technology were envisioned to provide the route to transhumanism. English evolutionary biologist Sir Julian Huxley, grandson of Thomas Henry Huxley, the great defender of Charles Darwin, and brother of author Aldous Huxley, who perhaps ironically wrote *Brave New World*, published an article in 1957 called "Transhumanism." Having coined the word and identified its "scientific possibilities," he concluded: "I believe in 'transhumanism': once there are enough people who can truly say that, the human species will be on the threshold of a new kind of existence . . . its real destiny."[4]

Nearly seventy years later, we're certainly not at that point yet, but suggestions for how to get there are more plentiful than they ever were before. Some I've touched on earlier in the book, like becoming a cyborg,

---

* Defined by the World Transhumanist Association as "a being whose basic capacities so radically exceed those of present-day humans as to no longer be unambiguously human by our current standards."

THE FUTURE OF US · 235

those individuals like Neil Harbisson and Kevin Warwick who have enhanced their bodies with arrays of electrodes or devices to extend their senses. There is now a small herd of people who have dabbled in a variety of prosthetics, some nearly practical, some more of a "look at me—I'm unique" project. Yes, they're pioneering in a sense, but still scratching the surface of transhumanism's ultimate goal.

Cold Comfort

There are also individuals who have placed their bets on cryonics, deep-freezing their bodies, or even just their brains, in the hopes that future technologies will bring them back to life. The American company Alcor, headquartered in Scottsdale, Arizona, offers to freeze your just-dead body, the sooner the better, by replacing your blood with antifreeze for $200,000 (technically life insurance) plus annual fees for stacking your body with others (upside down, four to a pod) at −196 degrees Celsius or −320.8 degrees Fahrenheit, the temperature of liquid nitrogen.[5] It's difficult to establish a precise number of those who have either undergone the procedure or have pledged to do so, although at the time of writing there are 191 "patients" (already frozen) and 1,353 "members" signed up for the process. Rumored to be among the latter group are some Silicon Valley types who seem disconcerted by the fact that they might die, including PayPal cofounder Peter Thiel and computer scientist and inventor Ray Kurzweil, author of *The Singularity Is Near*.*

Cryonics is questionable on so many levels: Does the biology even make sense? There is some evidence that some tissues can be frozen, thawed, and come back to life: wood frogs, for instance, freeze solid over the winter but revive in spring, although the temperatures are a mere few degrees below zero, not −196 degrees Celsius. It's also true that slices of rat brain have been treated with cryoprotectants and taken down to −130 degrees Celsius, then revived and apparently function normally.

---

* The "singularity" is Kurzweil's term for the moment when AI becomes superintelligent.

But whole brains? Never. And we're talking here about people who have died and then been transported to Alcor, with the inevitable decay over the course of the trip. This is why they suggest that if you're going to do it, it would be beneficial to live in Scottsdale. Also, a good argument can be made that it is uncertain when the necessary restorative technology will be available; and it will likely be centuries rather than decades in the future. What sort of world will welcome humans from the twenty-first century? How will those humans fare, given that they will be hopelessly out-of-date and clueless about their new environment?

The patients' successful comeback hinges on Alcor remaining solvent, not falling prey to the myriad ways a corporation can fail and not being irreparably damaged by natural disasters or even sabotaged.

An interesting option in the cryonics catalog is just having your head frozen. It's cheaper, only $80,000, and predicated on the idea that by the time technology actually exists, the brain will be all that's important. Robot bodies, arms, and legs can easily be attached to the human head. This idea, of course, is reminiscent of uploading the brain to AI: bodies are not necessary.

Is It Really All About the Brain?

Melanie Mitchell at the Santa Fe Institute argues that viewing the brain as completely independent from the body is a fallacy and that the idea of "embodied" cognition is ignored in both these schemes designed to defy death.

Embodied cognition asserts that our brains are not working in isolation, but rather that our thinking processes are strongly connected to the rest of the body; that reasoning, emotions, sensory input, motor controls are not just inputs and outputs to and from the brain but are integral parts of all mental activity. In this view, naturally the brain is important in thinking and awareness, but it's not alone.

By contrast, all significant development of AI has depended on the computational power of both hardware and software without any significant input from a body of any kind. While humanoid robots do interact with their bodies to control them, Mitchell argues that the type of feedback their computer brains receive is still very limited. Can we guarantee that illusions of body and limbs supplied to the uploaded brain will be enough to restore the whole brain?

Evidence for the close relationship between the way we think and express ourselves and our bodies can be found in common expressions like she's at the *top* of her game; don't let me *down*; darkness *pressed in* on all sides; that story was so *touching*; I dismissed that idea *out of hand*; and, my favorite from an article about the future of AI, "We *hold* the reins of the future in our hands." It's not unlike the list of commonsense (although admittedly ambiguous) headlines I wrote about earlier—just as AI would have trouble sorting them out, it might be puzzled by these embodied phrases when first confronted with them.

There's even evidence that thoughts and movements are directly connected. In one set of experiments, when volunteers were asked to think about the future, they swayed very slightly forward; thinking about the past caused them to sway backward!

Mitchell is convinced this involvement of the body is an important part of thinking that is overlooked by the developers of AI. "Nothing in our knowledge of psychology or neuroscience supports the possibility that 'pure rationality' is separable from the emotions and cultural biases that shape our cognition and our objectives," she says.[6]

Imagine the frozen head that's brought back to life two hundred years from now and finds, lacking a body, that it simply isn't the same person.

A scary thought. However, optimists, especially those who support extending and improving human life with superintelligent AI, remain resolute. Even though Hans Moravec wrote these words in the late 1980s, they continue to be enthusiastically embraced: "It is easy to imagine human thought freed from bondage to a mortal body—belief in an afterlife is common. But it is not necessary to adopt a mystical or religious stance to accept the possibility. Computers provide a model for even the most ardent mechanist."

Are We Living in a Computer Simulation?

> We are living in a computer programmed reality and the only clue we have to it is when some variable is changed and some alteration in our reality occurs, we would have the overwhelming impression that we are reliving the present, déjà vu.
>
> —Philip K. Dick, Metz, France, 1977

I'd argue that topic #1 in this chapter, consciousness, as resistant to understanding as it is, is amenable to scientific enquiry; topic #2, transhumanism, not so much; but then there's topic #3, posed here: "Are We Living in a Computer Simulation?" There's some science, or at least mathematics and statistics, that is appropriate to this question, but the premise is so far out that it's hard to take it seriously. Imagine this: the world we live in, the people you know—and most important, yourself—are not flesh-and-blood human beings, but computer-generated automata so detailed in our construction that we even have our conscious life. Like *The Sims*, only more high-end.

Who programmed us? For the moment let's assume that we've been programmed by the AI of the future, which certainly would be capable of doing it, perhaps at the whim of a future human. It could even be a future couch potato whiling away the time in self-amusement by doing things like creating this world. Regardless of identity, I'll refer to it as "the programmer." Some argue that the motivation might be nostalgia. In other words, there really was a twenty-first century, some time ago, and we are the copy of it. We have no way of knowing how much of what we're living is designed to approximate what actually happened. We're computer creations; we only know what the software creating us allows.

If you think that's nonsense, remember solipsism, the claim that there really is no way of knowing if there is *anything* outside of you and your thoughts; that it may be possible that there is nothing else other than what you imagine to exist: your friends, your children, the world simply created by you. It is a legitimate philosophical idea. How different is the idea that we're living in a computer simulation?

It's easy to denounce the idea and dismiss it, but let's entertain it if only until the end of the chapter. I mean, if author Philip K. Dick (*The Man in the High Castle, Do Androids Dream of Electric Sheep?*, *Minority Report*) thought we are in a computer simulation, who's to argue?

Imagine, it's the year 2123. Computers are scarily powerful. That future couch potato has chosen to run a computer game of life in the early twenty-first century. Maybe it's a history assignment in his or her school; maybe it's a way of doing recreational time travel. But the computer is set to re-create life as it's thought to have been back then. So the couch potato creates a simulation of breathtaking accuracy. Once it's set in motion, it can continue without interference from the programmer—if he or she wants. But if the couch potato wants to mix things up, if the created civilization is getting a little dull, then maybe tossing in a pandemic or something like that would be an option.

It Might Not Be Nonsense . . .

Think it'll never happen? Think it couldn't happen? It may sound crazy, but maybe it's not as crazy as you think. Philosopher Nick Bostrom of Oxford University, whom we have met before, came up with the so-called simulation argument in 2003,[7] and it's expressed by three possibilities:

1. Technological civilizations always go extinct before they reach the point where they can create such simulations.

2. If civilizations do acquire this ability, they have no interest in doing it.

3. You are almost certainly in a simulation.

Unless either 1 or 2 happens, 3 becomes inevitable. Phrased another way, either we disappear, or future humans don't run simulations, or we're in one. These principles have not gone uncriticized. Reinterpretation of the statistical arguments behind the conclusion of #3, that we're almost certainly in a simulation if 1 and 2 are not true, has prompted arguments that it may not be "almost certainly," but it might be hovering around half-certain, which still leaves the door pretty wide open. Some argue that while we have no idea what the probabilities really are, as soon as a simulation is actually created, then there's likely to be many more simulated individuals than real ones.

Elon Musk has advanced a simpler argument: given how quickly we've advanced from 2-D video games like *Pong* to incredibly complex, detailed, and realistic multiplayer video games—a few decades at most— then virtually any rate of continuing improvement will guarantee the existence of simulations. His guess is that the chance we're living a real existence is "one in billions." Of course, basing the creation of a simulation, where the characters believe their existence is real, on video games is neatly omitting the issue of how to make them conscious, given that we don't even know how consciousness works.

Jason Kehe, senior editor and culture critic at *Wired* magazine, put it nicely when he claimed that three things need to happen to establish a crackpot idea: "(1) its nonthreatening introduction to the masses, (2) its legitimization by experts, and (3) overwhelming evidence of its real-world effects."[8] In his opinion, those were, in order, the 1999 film *The Matrix*, Nick Bostrom's 2003 paper, and a mélange of cultural/political stuff, from Donald Trump's 2016 election as president to the 2017 Academy Awards. On a more serious note, Dartmouth College scientist Marcelo Gleiser argues that acknowledging that we're helpless players in a hypersophisticated computer game would give us license to abandon efforts to change things for the better, a return to the attitude, "It's God's will, what can I do?"[9] We're a long way from taking the simulation idea that seriously. The problem is that *actual* hard evidence doesn't exist, although some scientists are searching for it.

Where Is the Evidence?

By the way, you're excused if you think all of this is tongue-in-cheek, although by doing so you'd be rejecting the modern version of Plato's shadows in the cave, the cave residents thinking the shadows of people flickering on the cave walls were real, not one step removed from the real people casting them. But read on: Is there any way we can tell that we're living in a computer simulation?

The conjecture I like best is based on a limitation that even our tech-savvy descendants might face: in today's gaming or movie worlds, creating realistic scenes is energy-demanding and therefore expensive.

The short version of this argument is that when you look at even the most high-resolution display, it doesn't bother with details that only a physicist would care about. Cosmologist and theoretical physicist John Barrow of Cambridge University put it best: "When the Disney company makes a film that features the reflection of light from the surface of a lake, it does not use the laws of quantum electrodynamics and optics to compute the light scattering. That would require a stupendous amount of computing power and detail."[10]

Barrow argues that the creator of the simulation would use cheaper shortcuts that would be good enough for the casual observer and might only be revealed if someone took the time and energy to investigate the deep physics of the environments around us. It would be analogous to the way crowd conversation in movies is simulated by having actors repeat "rhubarb, rhubarb," "rabble, rabble," or "peas and carrots," over and over. But, like those nonsense repetitions, the visual analogue wouldn't stand up under close inspection. Is it fair to extrapolate this to a future where we'd expect that computers are much more powerful and would be able to create more richly detailed representations of reality, but would still be cost- and energy-conscious?

Barrow and others think they could face the same trade-offs as we do today, and therefore a search right now for inadequate detail might be worthwhile. Barrow's take is that however sophisticated the knowledge of the laws of nature was when the simulation began, inevitable discrepancies would start to crop up and the only solution available to the programmer would be to update, just as your laptop software requires periodic updating. Maybe, just maybe, we might notice that, say, a fundamental constant of nature has changed slightly. Or we might not, if the programmer makes the changes almost imperceptible and extending over eons of time—undetectable.

I confess to wondering why the programmer of our simulation would be so anxious to update the laws of physics here, when shortcomings in them are discovered in the real world, the one the programmer presumably inhabits. Is there a requirement for a simulation to represent the real world as accurately as possible? Why couldn't the programmer just let our simulation run off the rails?

Regardless, physicists working at the absolutely smallest scales, trying to simulate minuscule fragments of the universe, have found that yes, some of their simulations accurately represent something like the simplest atomic nucleus, but at the same time generate artifacts. The artifacts are the giveaway that it's a simulation. As the scientists continue to scale up, they think they might one day discover comparable artifacts that could reveal the much, much larger simulation.

Fouad Khan, a senior editor at the journal *Nature Energy*, proposed in *Scientific American* magazine that the fact that there's an upper limit in the universe, the speed of light, is proof that we live in a simulation, because computer-driven processes always have an upper limit, too: processor speed. It is a permanent artifact of computing.[11] So in his words: "The speed of light meets all the criteria of a hardware artifact . . . it is observed as a maximum limit, it is unexplainable by the physics of the universe, and it is absolute. The speed of light is a hardware artifact showing we live in a simulated universe."*

Again, a pretty cool approach but still a bit remote from daily life. There is unfortunately no straightforward common experience that could be construed as a revelation that this is all just a simulation. There is Philip K. Dick's suggestion that déjà vu is evidence of a change in a variable. It's not a bad idea to use déjà vu as an example of something otherworldly, or at least other than *this* worldly, because the standard explanations for déjà vu aren't very convincing. Yet if you're going to use it, you have to explain why experiences of déja vu happen less and less

---

* In the same article, Khan links the simulation to consciousness by arguing that consciousness is just a way of displaying our mental processes. To whom? The programmer.

often as we get older—are we only sensitive to adjustments to our simulated reality when we're young?

At the risk of misinterpreting Elon Musk, or venturing further than he might be willing to go (is that even possible?), he is associated with the idea that simulations must be more interesting than the world in which they were created, in the same way that video games and movies are highly condensed and selective versions of our world. If you accept that a world's simulations are always going to be more vivid than life in the world itself, and you further imagine that once created a simulation can create its own simulations, you have to conclude that the ultimate simulation is pretty fantastic, but the real world that kicked off the stack of simulations must be incredibly dreary.

Finally, so what? Actually, there isn't much about being part of a simulation that matters unless the person who programmed it tires of it, in which case we're doomed when he or she pulls the plug.

Motivated by this alarming possibility, there have even been suggestions as to how best to conduct ourselves if we have a high probability of being simulated beings in a simulated reality. Even though we really don't know the statistical probability that we are, wouldn't it be useful, maybe even important, to behave in a way that would maintain the interest of the programmer, so the simulation (our world) doesn't just get turned off because it's boring?

Best Behavior

Economist and author Robin Hanson has some recipes for survival in these circumstances. Imagine, he wrote, that simulations do tend to get turned off and in fact most don't last forever: then worrying about the future doesn't make much sense, but living for today does.*

To discourage simulation-ending thoughts, you might also want to make your surroundings and your actions as fascinating as you possibly

---

* Is there an irony here that worrying about the future isn't advisable, but the only reason that rule exists is that we *are* thinking about the future?

can, especially if our simulation is designed to highlight interesting people of the past. (The past being now, of course.) I'm fond of this particular piece of advice from Hanson, although I don't really see why our simulation has to be, in any way, some approximation of actual history. Why couldn't it just be an alternative history that might have kicked off, say, a billion years ago? In that case, ancient Rome might only have existed in our simulation, not in the programmers' real world. Or ancient Rome didn't even exist in our simulation, but instead the program running us has implanted that history in our minds.

How does Hanson think we could capture the attention of the future programmer and keep the finger off the button? "Be funny, outrageous, violent, sexy, strange, pathetic, heroic . . . in a word 'dramatic.' Being a martyr might even be a good thing for you, if that makes your story so compelling that other descendants will also want to simulate you."[12]

It's All Real—Isn't It?

This section began with the serious science of consciousness and ended with the absurd idea (at least to some) that we live in someone's (or something's) computer program. David Chalmers is the guy to tie it all together. A philosopher at New York University, he has been prominent in consciousness studies for decades, and in his latest book, *Reality+: Virtual Worlds and the Problems of Philosophy*, he turns his attention to the idea of the computer simulation (although his interest goes back to the fact that he consulted on the making of *The Matrix*).

Chalmers argues that simulated reality, digital reality, is equivalent to what we think of as the real thing and in fact can be as rewarding. If we're actually in a simulation, it *is* our reality, whether we are comfortable with that possibility or not. And while that's all uncertain, we already know that things are not the way they seem. Quantum mechanics is a very different view of reality than the world we perceive with our senses: that table you rest your elbows on is mostly empty space. We've accommodated to that notion if we think about it at all.

He writes, "If I've lived my whole life in a simulation, every flower I experience has been digital all along,"[13] tempting us to think, *Well, it's all right then, isn't it?* That digital flower is the thing that I've judged to be beautiful. However, if our simulation is based on a reality somewhere, we'll never know if the real flower is even better.

Acknowledgments

I am incredibly lucky to have a great relationship with Simon & Schuster Canada, including and especially president and publisher Kevin Hanson and fabulous editors like Nita Pronovost (aka Nita Prose), Sarah St. Pierre, and for this book, for the first time, Janie Yoon. How can an author resent Janie's suggestions in the margins when each begins "Hi Jay" and is of course the right move to make anyway? Somehow the stars aligned for Janie to be my editor and I couldn't be happier.

Trying to write sensibly about multiple futures cannot be a solo task, and I thank Andrew Hessel, Mark Poznansky, Marie Claire Arrieta, Shannon Falconer, Geoff Holmes, Niki Wilson, Tanya Roussy, Mark Redmond, Emily Grant, and Jeremie Harris for very helpful suggestions.

I have been writing books for twenty-five years and have always been fortunate to have friends and family to serve as surrogate audiences; this time is no exception. So thanks for your patience, Rachel, Paul, Amelia, Brendan, Max, Chanel, Grace, Everett, Finn, Olivia, Jasper, Alissa, all the Mosers, the Bindlestiffs, and the Band. A special shout-out to the Flatterheads.

Notes

Introduction: The Fog of the Future

1. For American biologist Paul Tesar's response to the creation of mouse embryos in a lab, see https://thedaily.case.edu/medicines-paul-tesar -discussed-stem-cells-and-reproductive-biology/.
2. Brett J. Kagan, Andy C. Kitchen, Nhi T. Tran, et al., "In Vitro Neurons Learn and Exhibit Sentience When Embodied in a Simulated Game-World," *Neuron* 110, no. 23, December 7, 2022, https://www.cell.com /neuron/fulltext/S0896-6273(22)00806-6.
3. "He Jankui and the World's First Gene-edited Babies," World Science Festival Shorts, https://www.youtube.com/watch?v=NRIs6xcqdEI.
4. Max Tegmark, *Life 3.0: Being Human in the Age of Artificial Intelligence* (New York: Knopf, 2017), Chapter 3, "The Near Future: Breakthroughs, Bugs, Laws, Weapons and Jobs."
5. Robert Reich, "The Failures of Big Tech and How to Fix It," *The Feedback Loop by Singularity*, Episode 66, August 8, 2022, https://podcasts.apple .com/fr/podcast/fbl66-rob-reich-the-failures-of-big-tech-how-to-fix-it/ id1468766317?i=1000575414457.

Part I: All About Us

Chapter 1: Our Future Selves

1. "The Sixth Finger," *The Outer Limits*, Season 1, Episode 5, https://www .dailymotion.com/video/x4jj142.
2. You may have seen the March of Progress before, but here's an interesting take on it: Kevin Blake, "On the Origins of 'The March of Progress,'" *Washington University ProSPER*, Washington State University, December 17, 2018, https://sites.wustl.edu/prosper/on-the-origins-of-the-march-of -progress/.

3. Kathleen McAuliffe, "The Shrinking Brain," *Discover*, August 1, 2011, https://www.discovermagazine.com/the-sciences/the-shrinking-brain.
4. "Anthropologists Find American Heads Are Getting Larger," *Science Daily*, May 30, 2012, https://www.sciencedaily.com/releases/2012/05 /120530115828.htm.
5. Richard H. Steckel and Joseph M. Prince, "Tallest in the World: Native Americans of the Great Plains in the Nineteenth Century," *American Economic Review* 91, no. 1 (March 2001): 287–94.
6. Timothy M. Waring and Zachary T. Wood, "Long-Term Gene-Culture Coevolution and the Human Evolutionary Transition," *Royal Society Publishing*, June 2, 2021, https://royalsocietypublishing.org/doi/10.1098 /rspb.2021.0538.
7. Michael Lynch, "Mutation and Human Exceptionalism: Our Future Genetic Load," *Genetics* 202, no. 3 (March 1, 2016): 869–75, https: //asu.pure.elsevier.com/en/publications/mutation-and-human -exceptionalism-our-future-genetic-load.
8. Mark Changizi, "Why Do We Have Ten Fingers?" *Science 2.0*, May 17, 2010, https://www.science20.com/mark_changizi/why_do_we_have_ten _fingers.
9. C. Mehring, M. Akselrod et al., "Augmented Manipulation Ability in Humans with Six-Fingered Hands," *Nature Communications*, June 3, 2019, https://www.nature.com/articles/s41467-019-10306-w.
10. For more information on the Third Thumb Project, see https://www .daniclodedesign.com/thethirdthumb.
11. Kerry Noble, Conrad Duncan, and Caroline Brogan, "Engineering Extra Limbs and Austerity Food Purchases: News from the College," *Imperial College London News*, March 18, 2022, https://www.imperial.ac.uk/news /234875/engineering-extra-limbs-austerity-food-purchases/.

Chapter 2: Restoration Hardware

1. Jacqueline Finch, "The Ancient Origins of Prosthetic Medicine," *Lancet*, February 12, 2011, https://www.thelancet.com/journals/lancet/article/ PIIS0140-6736(11)60190-6/fulltext.
2. Thomas Morris, "Miracle in Capetown," *Lancet*, December 2, 2017, https: //www.thelancet.com/journals/lancet/article/PIIS0140-6736(17)32912 -4/fulltext.
3. Matthias Fuchs et al., "Does the Heart Transplant Have a Future?"

European Journal of Cardio-Thoracic Surgery, June 1, 2019, https://pubmed.ncbi.nlm.nih.gov/31106338/.

4. Syed Shahyan Bahktiyar et al., "Survival on the Heart Transplant Waiting List," *JAMA Cardiology,* August 12, 2020, https://jamanetwork.com/journals/jamacardiology/fullarticle/2769179.

5. For more information on BiVACOR, see https://bivacor.com/.

6. Wei Wang et al., "First Pig-to-Human Heart Transplant," *Innovation,* March 29, 2022, https://www.ncbi.nlm.nih.gov/pmc/articles/PMC8961469/.

7. "Mind-Controlled Prosthetic," University of Newcastle, Australia, https://www.youtube.com/watch?v=_-3lGMjnGi0.

8. Hugh Herr's TED Talk: https://www.ted.com/talks/hugh_herr_the_new_bionics_that_let_us_run_climb_and_dance?language=en#t-880620.

9. Stéphanie Roy and Samuel Gatien, "Regeneration in Axolotls," *Experimental Gerontology* 43, no. 11 (November 2008): 968–73, https://www.sciencedirect.com/science/article/abs/pii/S0531556508002933.

10. Mike Silver, "Scientists Regrow Frog's Lost Leg," *TuftsNow,* January 26, 2022, https://now.tufts.edu/2022/01/26/scientists-regrow-frogs-lost-leg#:~:text=On%20adult%20frogs%2C%20which%20are,stump%20for%20just%2024%20hours.

11. Nirosha J. Murugan et al., "Acute Multidrug Delivery Via a Wearable Bioreactor Facilitates Long-Term Limb Regeneration and Functional Recovery in Adult *Xenopus laevis,*" *Science Advances,* January 26, 2022, https://www.science.org/doi/10.1126/sciadv.abj2164.

Chapter 3: Cyborgs

1. For the opening and closing theme of *The Six Million Dollar Man,* see https://www.youtube.com/watch?v=0CPJ-AbCsT8.

2. Hugh Herr, "How We'll Become Cyborgs and Extend Human Potential," TED Talk, May 30, 2018, https://www.youtube.com/watch?v=PLk8Pm_XBJE.

3. Manfred E. Clynes and Nathan S. Kline, "Cyborgs and Space," *New York Times,* 1997, reprinted with permission from *Astronautics,* September 1960, https://archive.nytimes.com/www.nytimes.com/library/cyber/surf/022697surf-cyborg.html.

4. Dominique Sisley, "Why This Artist Got an Antenna Implanted in His Skull," *Dazed,* May 12, 2016, https://www.dazeddigital.com/artsandculture/article/31102/1/why-this-artist-got-an-antenna-implanted-in-his-skull.

5. Jay Ingram, *Daily Planet: The Ultimate Book of Everyday Science* (Toronto: Penguin Canada, 2010), 235.

6. Daniela Cerqui et al., "Re-designing Humankind: The Rise of Cyborgs, a Desirable Goal?" *Philosophy and Design*, January 2008, 185–95, https://www.researchgate.net/publication/226089874_Re-Designing _Humankind_The_Rise_of_Cyborgs_a_Desirable_Goal.

7. Linxing Jiang et al., "BrainNet: A Multi-Person, Brain-to-Brain Interface for Direct Collaboration Between Brains," *Nature*, April 16, 2019, https: //www.nature.com/articles/s41598-019-41895-7#:~:text=We% 20present%20BrainNet%20which%2C%20to,interface%20for%20 collaborative%20problem%20solving.

8. Peter Dockrill, "Brain Implant Translates Paralyzed Man's Thoughts into Text with 94% Accuracy," *Science Alert*, November 8, 2021, https:// www.sciencealert.com/brain-implant-enables-paralyzed-man-to -communicate-thoughts-via-imaginary-handwriting.

9. Moises Velasquez-Manoff, "The Brain Implants That Could Change Humanity," *New York Times*, August 28, 2020, https://www.nytimes .com/2020/08/28/opinion/sunday/brain-machine-artificial-intelligence .html?referringSource=articleShare%C2%A0.

10. Antonio Regalado, "An ALS Patient Set a Record for Communicating via a Brain Implant: 62 Words per Minute," *MIT Technology Review*, January 24, 2023, https://www.technologyreview.com/2023/01/24 /1067226/an-als-patient-set-a-record-for-communicating-via-a-brain -implant-62-words-per-minute/.

11. Mark Bergen, "Here's the Cyborg Tech That Elon Musk Says He'll Do If No One Else Does," *Vox*, June 2, 2016, https://www.vox.com/2016/6/2 /11837544/elon-musk-neural-lace.

12. Rupert Neate, "Elon Musk's Brain Chip Firm Neuralink Lines Up Clinical Trials in Humans," *Guardian*, January 20, 2022, https://www.theguardian .com/technology/2022/jan/20/elon-musk-brain-chip-firm-neuralink -lines-up-clinical-trials-in-humans.

13. Dan Robitzski, "Expert: Neuralink Could Sell Your Private Thoughts to the 'Highest Bidder,'" *Neoscope*, April 14, 2021, https://futurism.com/ neoscope/expert-neuralink-sell-private-thoughts-highest-bidder.

14. Hugh Herr, "How We'll Become Cyborgs and Extend Human Potential," TED Talk, May 30, 2018, https://www.youtube.com/watch?v=PLk8Pm _XBJE.

Chapter 4: It's in Our DNA

1. Sergey Nurk et al., "The Complete Sequence of a Human Genome," *Science*, March 31, 2021, https://www.science.org/doi/10.1126/science.abj6987.

2. For more on the Human Genome Project, visit the National Human Genome Institute: https://www.genome.gov/human-genome-project.

3. John Cryan, "Feed Your Microbes, Nurture Your Mind," TEDx Talk, July 11, 2017, https://www.youtube.com/watch?v=vKxomLM7SVc.

4. Helen E. Vuong et al., "The Maternal Microbiome Modulates Fetal Neurodevelopment in Mice," *Nature*, October 2020, 281–86, https://pubmed.ncbi.nlm.nih.gov/32968276/.

5. Jessica Bayes, Janet Schloss, and David Sibbritt, "Effects of Polyphenols in a Mediterranean Diet on Symptoms of Depression: A Systematic Literature Review," *Advances in Nutrition* 11, no. 3 (May 2020): 602–15, https://www.ncbi.nlm.nih.gov/pmc/articles/PMC7231605/.

6. "Microbiota Therapy: The Dawn of a New Age in Modern Medicine," *Healthy Daily*, November 17, 2021, https://thehealthydaily.com/2021/11/17/microbiota-therapy-the-dawn-of-a-new-age-in-modern-medicine/.

7. Antonio Regalado, "The Creator of the CRISPR Babies Has Been Released from a Chinese Prison," *MIT Technology Review*, April 4, 2022, https://www.technologyreview.com/2022/04/04/1048829/he-jiankui-prison-free-crispr-babies/.

8. For more on AlphaFold, see https://www.deepmind.com/research/highlighted-research/alphafold.

9. "Stopping Malaria in Its Tracks," *Unfolded*, DeepMind, October 13, 2022, https://unfolded.deepmind.com/stories/matthew-higgins-is-unlocking-a-new-path-to-stop-malaria-in-its-tracks.

10. For more information on the Centre for Enzyme Innovation at the University of Portsmouth, visit https://www.port.ac.uk/research/research-centres-and-groups/centre-for-enzyme-innovation.

11. Ewen Callaway, "AlphaFold's New Rival? Meta AI Predicts Shape of 600 Million Proteins," *Nature*, November 1, 2022, https://www.nature.com/articles/d41586-022-03539-1.

Chapter 5: The Good Long Life

1. Nikolay Zak, "Evidence that Jeanne Calment Died in 1934—Not 1997," *Rejuvenation Research* 22, no. 1 (2019), https://scholar.google.ca/scholar_url?url=https://www.liebertpub.com/doi/pdf/10.1089/rej.2018.2167%

3Ffbclid%3DIwAR0gjUI1SeOHeTKv5wSmdYNjn--ArQIN9LFo8Gqff0
ASiOGywmqUQTUsGzA%26&hl=en&sa=X&ei=umr-Y_qdBYWY
ywTYn4T4Ag&scisig=AAGBfm1iLRveu0wH-TnOhQL6xaQAgAKr0
Q&oi=scholarr.

2. Kyung-Jin Min, Cheol-Koo Lee, and Han-Nam Park, "The Lifespan of Korean Eunuchs," *Current Biology* 22, no. 18 (September 25, 2022): R792–R793, https://www.sciencedirect.com/science/article/pii/S0960982212007129.

3. American Association for the Advancement of Science News Release, "Science's Breakthrough of the Year: Uncovering 'Ardi,'" *EurekAlert*, December 17, 2019, https://www.eurekalert.org/news-releases/816289.

4. G. A. Lindeboom, "The Story of a Blood Transfusion to a Pope," *Journal of the History of Medicine and Allied Sciences* 9, no. 4 (October 1, 1954): 455–59, https://academic.oup.com/jhmas/article-abstract/IX/4/455/932051.

5. Annalee Armstrong, "Anti-Aging Foundation SENS Fire de Grey after Allegations He Interfered with Investigation into His Conduct," *Fierce Biotech*, August 23, 2021, https://www.fiercebiotech.com/biotech/anti-aging-foundation-sens-turfs-de-grey-after-allegations-he-interfered-investigation-into.

6. Alvin Powell, "Has First Person to Live to Be 150 Been Born?" *Harvard Gazette*, January 30, 2023, https://news.harvard.edu/gazette/story/2023/01/has-first-person-to-live-to-be-150-been-born/.

7. Peter H. Diamandis, MD, "Dangerous Ideas from Ray Kurzweil," *Singularity Hub*, November 10, 2017, https://singularityhub.com/2017/11/10/3-dangerous-ideas-from-ray-kurzweil/#:~:text=This%20will%20keep%20accelerating%20as,will%20hit%20longevity%20escape%20velocity.%E2%80%9D

8. For more information on Longevity House, visit https://www.longevityhouse.ca/.

Part II: What Will We Eat?

Chapter 6: Food Fight

1. For more information on the EAT-Lancet Commission on Food, Planet, Health, visit https://eatforum.org/eat-lancet-commission/.

2. Paul R. Ehrlich and Anne H. Ehrlich, "The Population Bomb Revisited," *Electronic Journal of Sustainable Development* 1, no. 3 (2009): 63–71, https://scholar.google.ca/scholar_url?url=https://sites01.lsu.edu/faculty /kharms/wp-content/uploads/sites/23/2017/04/Ehrlich_Ehrlich_2009 _EJSD_ThePopulationBombRevisited.pdf&hl=en&sa=X&ei=hHH -Y8T3Epn4yATzjJS4CA&scisig=AAGBfm3UFkSq40SMabQyHmEQBz MWYTA7FA&oi=scholarr.

3. United Nations, Sixty-Fourth General Assembly, Second Committee, "Food Production Must Double by 2050 to Meet Demand from World's Growing Population, Innovative Strategies Needed to Combat Hunger, Experts Tell Second Committee," October 9, 2009, https://press.un.org/ en/2009/gaef3242.doc.htm#:~:text=Food%20production%20must%20 double%20by%202050%20to%20meet%20the%20demand,a%20 panel%20discussion%20on%20%E2%80%9CNew.

4. Erica M. Goss et al., "The Irish Potato Famine Pathogen *Phytophthora infestans* Originated in Central Mexico Rather than the Andes," *PNAS*, June 2, 2014, https://www.pnas.org/doi/10.1073/pnas.1401884111#:~: text=Phytophthora%20infestans%20is%20a%20destructive,potato%20 pathogen%20to%20manage%20worldwide.

5. Tadessa Daba, "New African Potato Resists the Same Disease That Caused the Irish Potato Famine," *ScienceAlert*, December 13, 2020, https: //www.sciencealert.com/scientists-create-a-new-potato-with-complete -resistance-to-a-disastrous-disease.

6. George Monbiot, "Towering Lunacy," *Guardian*, August 17, 2010, https:// www.monbiot.com/2010/08/16/towering-lunacy/.

7. Senthold Asseng et al., "Wheat Yield Potential in Controlled-Environment Vertical Farms," *PNAS* 17, no. 32 (August 11, 2020): 18895, https://www .ncbi.nlm.nih.gov/pmc/articles/PMC7430987/#.

Chapter 7: The Issue Is Meat

1. Frederick Edwin Smith, 1st Earl of Birkenhead, *The World in 2030 A.D.* (London: Hodder & Stoughton, 1930), 19.

2. P. Sans and P. Combris, "World Meat Consumption Patterns: An Overview of the Last 50 Years (1961–2011)," *Meat Science* 109 (November 2015): 106–11, https://www.sciencedirect.com/science/article/abs/pii/ S0309174015300115.

3. Martin C. Parlasca and Matin Qaim, "Meat Consumption and Sustaina-

bility," *Annual Reviews*, 2022, 17–41, https://www.annualreviews.org/doi/pdf/10.1146/annurev-resource-111820-032340.

4. Food and Agriculture Organization of the United Nations, "Key Facts and Findings," https://www.fao.org/news/story/en/item/197623/icode/.

5. Michael B. Eisen and Patrick O. Brown, "Rapid Global Phaseout of Animal Agriculture Has the Potential to Stabilize Greenhouse Gas Levels for 30 Years and Offset 68 Percent of CO_2 Emissions This Century," *Plos Climate*, February 1, 2022, https://journals.plos.org/climate/article?id=10.1371/journal.pclm.0000010.

6. United Nations Climate Change, "Impossible Foods: Creating Plant-Based Alternatives for Meat: Singapore, Hong Kong, USA, Macau," 2019, https://unfccc.int/climate-action/momentum-for-change/planetary-health/impossible-foods#:~:text=Helping%20the%20Plane&text=Compared%20to%20this%20beef%2C%20the,pollution%20to%20fresh%20water%20ecosystems.

7. Agrifood Analytics Lab, Research, https://www.dal.ca/sites/agri-food/research/animal-vs-vegetable-protein-prices.html.

8. Patrick Greenfield, "'Let's Get Rid of Friggin' Cows' Says Creator of Plant-Based 'Bleeding Burger,'" *Guardian*, January 8, 2021, https://www.theguardian.com/environment/2021/jan/08/lets-get-rid-of-friggin-cows-why-one-food-ceo-says-its-game-over-for-meat-aoe.

9. Amar Toor, "World's First Lab-Grown Burger Unveiled at Public Tasting," *Verge*, August 5, 2013, https://www.theverge.com/2013/8/5/4589744/cultured-beef-burger-public-tasting-mark-post-sergey-brin.

10. John Lynch and Raymond Pierrehumbert, "Climate Impacts of Cultured Meat and Beef Cattle," *Frontiers Sustainable Food Systems* 3 (2019), https://www.frontiersin.org/articles/10.3389/fsufs.2019.00005/full.

11. Gregory S. Okin, "Environmental Impacts of Food Consumption by Dogs and Cats," *Plos One*, August 2, 2017, https://journals.plos.org/plosone/article?id=10.1371/journal.pone.0181301.

12. Amy Webb, "Welcome to the 'Synthetic' Meatspace," *Wired*, December 1, 2021, https://www.wired.com/story/synthetic-meatspace-food-health/?mbid=social_twitter&utm_brand=wired&utm_medium=social&utm_social-type=owned&utm_source=twitter.

13. Greenfield, "'Let's Get Rid of Friggin' Cows.'"

14. Chyrantus M. Tangus et al., "Edible Insect Farming as an Emerging and Profitable Enterprise in East Africa," *Current Opinion in Insect Science*

48 (December 2021): 64–71, https://www.sciencedirect.com/science /article/pii/S2214574521001073#:~:text=Highlights&text=Insect%20 farming%20is%20a%20novel,and%20income%20in%20East%20Africa .&text=More%20than%2080%25%20of%20feed,their%20livestock%20 and%20fish%20feeds.&text=Over%2065%25%20of%20those%20 consuming,flour%20to%20whole%20insect%20products.

15. Klaus W. Lange and Yukiko Nakamura, "Edible Insects as Future Food: Chances and Challenges," *Journal of Future Foods* 1, no. 1 (September 2021): 38–46, https://www.sciencedirect.com/science/article/pii/S2772 566921000033.

16. Carolyn Beans, "Salted Ants. Ground Crickets. Why You Should Try Edible Insects," *Washington Post*, November 27, 2022, https://www .washingtonpost.com/health/2022/11/27/eating-insects-good-for-you/.

17. Paul Schattenberg, "Bug Banquet Makes a Unique Culinary Experience," *AgriLife Today*, November 10, 2015, https://agrilifetoday.tamu.edu/2015 /11/10/bug-banquet/.

18. Shawna Williams, "Spilling the Tea: Insect DNA Shows Up in World's Top Beverage," *Scientist*, June 14, 2022, https://www.the-scientist.com/news -opinion/spilling-the-tea-insect-dna-shows-up-in-world-s-top-beverage -70131.

Chapter 8: The Banana and the Broiler

1. James Dale et al., "Transgenic Cavendish Bananas with Resistance to Fusarium Wilt Tropical Race 4," *Nature Communications*, November 14, 2017, https://www.nature.com/articles/s41467-017-01670-6.

2. Rebecca Cohen, "Global Issues for Breakfast: The Banana Industry and Its Problems FAQ (Cohen Mix)," *Science Creative Quarterly*, June 12, 2009, https://www.scq.ubc.ca/global-issues-for-breakfast-the-banana-industry -and-its-problems-faq-cohen-mix/.

3. H. L. Shrader, "The Chicken-of-Tomorrow Program; Its Influence on 'Meat-Type' Poultry Production," *Poultry Science* 31, no. 1 (January 1, 1952): 3–10, https://www.sciencedirect.com/science/article/pii/S0032579 119513013.

4. Evie Liu, "Why Chickens Are Twice as Big Today as They Were 60 Years Ago," *MarketWatch*, January 6, 2017, https://www.marketwatch.com /story/why-chickens-are-twice-as-big-today-as-they-were-60-years-ago -2017-01-06.

5. Carys E. Bennett et al., "The Broiler Chicken as a Signal of a Human Reconfigured Biosphere," *Royal Society Publishing*, December 12, 2018, https://royalsocietypublishing.org/doi/10.1098/rsos.180325#:~:text=Broiler%20chickens%2C%20now%20unable%20to,reconfiguration%20of%20the%20Earth's%20biosphere.

6. Ekrem Misimi et al., "GRIBBOT—Robotic 3D Vision-Guided Harvesting of Chicken Fillets," *Computers and Electronics in Agriculture* 121 (February 2016): 84–100, https://www.sciencedirect.com/science/article/abs/pii/S0168169915003701.

7. Aaron Souppouris, "Robot Uses 3D Imaging and Force Feedback to Debone Chickens," *Verge*, May 31, 2012, https://www.theverge.com/2012/5/31/3054316/robot-deboning-chicken-3d-imaging-force-feedback-video.

Part III: Where Will We Live and How Will We Get There?

Chapter 9: Planes, Trains, and Something Elon Musk Said

1. International Energy Agency, *Report on Aviation*, September 2022, https://www.iea.org/reports/aviation.

2. Ulrike Bergström, "One Percent of the World's Population Accounts for More than Half of Flying Emissions," *Lund University News*, November 19, 2020, https://www.lunduniversity.lu.se/article/one-percent-worlds-population-accounts-more-half-flying-emissions#:~:text=The%20most%20frequent%20fliers%20constitute,individual%20users%20of%20private%20aircrafts.

3. For more information on Boom's XB-1, visit https://boomsupersonic.com/overture.

4. For more information on ZeroAvia, visit https://www.zeroavia.com/do228-first-flight.

5. For more information on Hybrid Air Vehicles's Airlander 10, visit https://www.hybridairvehicles.com/our-aircraft/airlander-10/.

6. Aitana Villegas, "China Tests the World's Fastest Train," *Diario AS*, October 21, 2022, https://en.as.com/latest_news/china-tests-the-worlds-fastest-train-n/.

7. Elon Musk, "Hyperloop Alpha," Tesla, 2012, https://www.tesla.com/sites/default/files/blog_images/hyperloop-alpha.pdf.

8. Ibid.

9. Paris Marx's Twitter thread on Elon Musk's hyperloop idea: https://twitter.com/parismarx/status/1557707438786330629.

10. Lauren Smiley, "'I'm the Operator': The Aftermath of a Self-Driving Tragedy," *Wired*, March 8, 2022, https://www.wired.com/story/uber-self-driving-car-fatal-crash/#:~:text=In%202018%2C%20an%20Uber%20autonomous,behind%20the%20wheel%20finally%20speaks.&text=Rafaela%20Vasquez%20liked%20to%20work,had%20her%20reasons%20to%20distrust.

11. P. A. Hancock, Illah Nourbakhsh, and Jack Steward, "On the Future of Transportation in an Era of Automated and Autonomous Vehicles," *Psychological and Cognitive Sciences* 116, no. 16 (January 14, 2019): 7684–91, https://www.pnas.org/doi/abs/10.1073/pnas.1805770115.

12. For more information on the Moral Machine, visit https://www.moralmachine.net/.

13. "History of the Flying Car, Part 2: The Curtiss Autoplane," Impact Lab, https://www.impactlab.com/2010/08/12/history-of-the-flying-car-part-two-the-curtiss-autoplane/.

14. "Waterman Aerobile," National Air and Space Museum, Collection Objects, https://airandspace.si.edu/collection-objects/waterman-aerobile/nasm_A19610156000.

15. For more information on the Aerocar, visit https://en.wikipedia.org/wiki/Aerocar.

16. Gideon Lichfield, "When Will We Have Flying Cars? Maybe Sooner than You Think," *MIT Technology Review*, February 13, 2019, https://www.technologyreview.com/2019/02/13/137462/when-will-we-have-flying-cars-maybe-sooner-than-you-think/.

17. Joby Aviation bought what was called "Uber Elevate," but Uber maintains a financial interest and the two companies plan to share apps, so you can use it to book either a car or a flight. Joby Aviation has a nice video of their flying car (which looks a lot like an airplane) on their website: https://www.jobyaviation.com/.

18. Oliver Milman, "NASA Leads Push for Electric Planes in Next Frontier of Cutting Emissions," *Guardian*, May 18, 2021.

19. Kevin DeGood, "Flying Cars Will Undermine Democracy and the Environment," Center for American Progress, May 28, 2020, https://www.americanprogress.org/article/flying-cars-will-undermine-democracy-environment/.

20. Ben Sampson, "Q&A: Parimal Kopardeker, Director of the NASA Aeronautics Research Institute," Aerospace Testing International, December 14, 2022, https://www.aerospacetestinginternational.com/features/qa-parimal -kopardekar-director-of-the-nasa-aeronautics-research-institute.html.
21. Adrienne Bernhard, "The Flying Car Is Here—and It Could Change the World," *BBC Future Inc.,* November 11, 2020, https://www.bbc.com/future /article/20201111-the-flying-car-is-here-vtols-jetpacks-and-air-taxis.

Chapter 10: Where Will We Live?

1. Karrie Jacobs, "Toronto Wants to Kill the Smart City Forever," *MIT Technology Review*, June 29, 2022, https://www.technologyreview.com /2022/06/29/1054005/toronto-kill-the-smart-city/.
2. Ibid.
3. Joseph Goodman, Melissa Laube, and Judith Schwenk, "Curitiba Bus System Is Model for Rapid Transit," *Moving the Movement for Transportation Justice* 12, no. 1 (Spring 2007), posted on *Reimagine RP&E Journal* website, https://www.reimaginerpe.org/curitiba-bus-system.
4. "The City of Chattanooga Digital Twin," City of Chattanooga, Tennessee, https://ulidigitalmarketing.blob.core.windows.net/ulidcnc/sites/5/2020 /03/COMSTOCK.pdf.
5. "What Can Today's Designers Learn from Nature?" *TED Radio Hour*, NPR, May 20, 2016, https://www.npr.org/transcripts/478566114.
6. Tokollo Matsabu, "America's Metropolises Are Becoming 'Sponge Cities' to Deal with Flooding, Rain, and More—This Is How It Works," *The Cool Down*, December 15, 2022, https://www.thecooldown.com/green-tech/ what-is-sponge-city-concept-pittsburgh/.
7. "How a Kingfisher Helped Reshape Japan's Bullet Train," BBC News, March 26, 2019, https://www.bbc.com/news/av/science-environ ment-47673287.
8. There's a good video of a kingfisher diving at https://asknature .org/innovation/high-speed-train-inspired-by-the-kingfisher/. But look at it closely: it's clear that it opens its mouth *before* hitting the water. So much for the aerodynamic effects of its closed bill! That goes against the theory and the redesign of the Shinkansen. Yet the train's specs improved, so what's going on?
9. Lee Billings, "The Termite and the Architect," *Nautilus*, December 13, 2013, https://nautil.us/the-termite-and-the-architect-234706/.

10. Roger S. Ulrich, "Natural Versus Urban Scenes: Some Psychophysiological Effects," *Environment and Behavior* 5, no. 13 (September 1981): 523–56, https://www.researchgate.net/publication/249623753_Natural_Versus _Urban_Scenes_Some_Psychophysiological_Effects.

11. Mathew P. White et al., "Spending at Least 120 Minutes a Week in Nature Is Associated with Good Health and Wellbeing," *Scientific Reports* 9, Article 7730 (June 13, 2019), https://www.nature.com/articles/s41598 -019-44097-3.

12. George Mackerron and Susan Mourato, "Happiness Is Greater in Natural Environments," *Global Environmental Change*, August 3, 2013, https:// eprints.lse.ac.uk/49376/1/Mourato_Happiness_greater_natural_2013.pdf.

13. "The Courtyard of the Future at Strassvej by BOGL," Landezine International Landscape Award, https://landezine-award.com/the -courtyard-of-the-future-at-straussvej/.

14. Masahiro Horiuchi et al., "Impact of Viewing vs. Not Viewing a Real Forest on Physiological and Psychological Responses in the Same Setting," *International Journal of Environmental Research and Public Health* 11, no. 10 (October 20, 2014): 10883–901.

15. Qing Li et al., "Phytoncides (Wood Essential Oils) Induce Human Natural Killer Cell Activity," *Immunopharmacology and Immunotoxicology* 28, no. 2 (2006): 319–33.

16. In *Why Look at Animals*, published in 1977, art critic John Berger described zoo animals as "a living monument to their own disappearance."

17. Felicia Keesing and Richard S. Ostfield, "Impacts of Biodiversity and Biodiversity Loss on Zoonotic Diseases," *PNAS* 118, no. 17 (April 5, 2021), https://www.pnas.org/doi/10.1073/pnas.2023540118.

18. Tim S. Doherty et al., "Invasive Predators and Global Biodiversity Loss," *PNAS* 113, no. 40 (September 16, 2016): 11261–65, https://www.pnas.org /doi/10.1073/pnas.1602480113.

19. World Wildlife Fund, *Living Planet Report 2022*, https://wwf.ca/wp -content/uploads/2022/10/lpr_2022_full_report_en.pdf.

20. Caspar A. Hallmann et al., "More than 75 Percent Decline Over 27 Years in Total Flying Insect Biomass in Protected Areas," *Plos One*, October 18, 2017, https://journals.plos.org/plosone/article?id=10.1371/journal.pone .0185809.

21. Jamie Rappaport Clark, "We Were Wrong About Wolves, Here's Why," Defenders of Wildlife, March 21, 2020, https://defenders.org/blog/2020/

03 /we -were -wrong -about -wolves -heres -why#:~:text=Wolf%20 reintroduction%20caused%20unanticipated%20change,%2C%20 eagles%2C%20foxes%20and%20badgers.

22. Michelle Ma, "Cougars Could Save Lives by Lowering Vehicle Collisions with Deer," *UW News*, University of Washington, July 14, 2016, https:// www.washington.edu/news/2016/07/14/cougars-could-save-lives-by -lowering-vehicle-collisions-with-deer/.

23. Lorraine Boissoneault, "When the Nazis Tried to Bring Animals Back from Extinction," *Smithsonian Magazine*, March 31, 2017, https://www .smithsonianmag.com/history/when-nazis-tried-bring-animals-back -extinction-180962739/.

24. "This Is the Only Animal in the World that Went Extinct Twice," *Earthly Mission*, https://earthlymission.com/bucardo-pyrenean-ibex -cloning-extinct-species-science/#:~:text=The%20Pyrenean%20ibex %20was%20the%20first%20animal%20to%20be%20resurrected%20 from%20extinction.

25. Paul Ehrlich and Anne Ehrlich, "The Case Against De-Extinction: It's a Fascinating but Dumb Idea," *Yale Environment 360*, January 13, 2014, https://e360.yale.edu/features/the_case_against_de-extinction_its_a _fascinating_but_dumb_idea.

26. For more information on Pleistocene Park, visit https://pleistocenepark.ru/.

27. Michael Greshko, "Mammoth-Elephant Hybrids Could Be Created Within the Decade. Should They Be?" *National Geographic*, September 14, 2021, https://www.nationalgeographic.co.uk/science-and-technology /2021/09/mammoth-elephant-hybrids-could-be-created-within-the -decade-should-they-be.

28. Brandon Specktor, "The CIA Wants to Bring Woolly Mammoths Back from Extinction," *Live Science*, October 13, 2022, https://www.livescience .com/cia-wooly-mammoth-de-extinction.

29. To learn more about the passenger pigeon, visit https://www.audubon.org /birds-of-america/passenger-pigeon#:~:text=The%20air%20was%20 literally%20filled,lull%20my%20senses%20to%20repose.

30. For more on the Passenger Pigeon Project, visit https://reviverestore.org/ about-the-passenger-pigeon/.

31. Ben Novak, "How to Bring Passenger Pigeons All the Way Back," TEDx DeExtinction, April 1, 2013, https://www.youtube.com/watch?v =rUoSjgZCXhc (at the 2:47 mark).

32. Mark Zuckerberg, "Founders Letter, 2021," Meta, October 28, 2021, https://about.fb.com/news/2021/10/founders-letter/.

33. If you doubt this could happen, check out this YouTube video of Epic Games' Unreal Engines animation: https://www.youtube.com/watch?v=d1ZnM7CH-v4. Epic Games is targeting the metaverse.

34. Gerhard Reese, Elias Kohler, and Claudia Menzel, "Restore or Get Restored: The Effect of Control on Stress Reduction and Restoration in Virtual Settings," *Sustainability* 13, no. 4 (February 12, 2021), https://doi.org/10.3390/su13041995.

35. Rachel Becker, "Scott Kelly's Year in Space Highlights Risk to DNA and Brains," *Verge*, April 11, 2019, https://www.theverge.com/2019/4/11/18306525/scott-mark-kelly-twins-year-international-space-station-nasa-dna-genes-health.

36. Gerard K. O'Neill, "The Colonization of Space," *Physics Today* 27, no. 9 (September 1974): 32–40.

37. "Orbital Space Settlements," National Space Society, https://space.nss.org/orbital-space-settlements/.

38. Daniel Terdiman, "Elon Musk at SXSW: 'I'd Like to Die on Mars, Just Not on Impact,'" *CNET*, March 9, 2013, https://www.cnet.com/culture/elon-musk-at-sxsw-id-like-to-die-on-mars-just-not-on-impact/.

Chapter 11: Tech for the Planet

1. Isobel Asher Hamilton, "Jeff Bezos Says Space Travel Is Essential Because We Are 'in the Process of Destroying This Planet,'" *Business Insider*, July 18, 2019, https://www.businessinsider.com/jeff-bezos-space-travel-essential-because-destroying-planet-2019-7.

2. Jakub Stachurski, "Reducing Carbon Emissions One Tree at a Time," *dcbel*, June 14, 2021, https://www.dcbel.energy/blog/2021/06/14/reducing-carbon-emissions-one-tree-at-a-time/.

3. Zayna Syed, "Can a Mechanical 'Tree' Slow Climate Change? An ASU Researcher Built One to Find Out," *AZCentral*, April 22, 2022, https://www.azcentral.com/story/news/local/arizona-environment/2022/04/22/asu-researcher-builds-mechanical-tree-capture-carbon-dioxide/7398671001/.

4. "Carbon Capture," Center for Climate and Energy Solutions, https://www.c2es.org/content/carbon-capture/#:~:text=Carbon%20capture%20can%20achieve%2014,decarbonization%20in%20the%20industrial%20sector.

5. "Climeworks Makes History with World's First Commercial Direct Air Capture Plant," Climeworks News, May 31, 2021, https://climeworks.com/news/today-climeworks-is-unveiling-its-proudest-achievement.

6. "Carbon Engineering Begins Work on Supporting Multi-Million Tonne Direct Air Capture Facilities in South Texas," Climeworks News & Updates, October 31, 2022, https://carbonengineering.com/news-updates/multi-million-tonne-south-texas/.

7. I. R. Kivi et al., "Multi-Layered Systems for the Geologic Storage of CO_2 at the Gigatonne Scale," Geophysical Research Letters 49, no. 24 (December 2022).

8. For more information on 1PointFive and their low-carbon fuels, visit https://www.1pointfive.com/air-to-fuels.

9. In-person communication between the author and Geoff Holmes, business development manager at Carbon Engineering, September 22, 2022.

10. Kevin Anderson and Glen Peters, "The Trouble with Negative Emissions," Science 354, no. 6309 (October 14, 2016): 182–83, https://www.science.org/doi/10.1126/science.aah4567.

11. For more information on stratospheric aerosol injection, visit https://en.wikipedia.org/wiki/Stratospheric_aerosol_injection.

12. For more information on space mirrors, visit https://en.wikipedia.org/wiki/Space_mirror_(climate_engineering).

13. Jennie C. Stephens et al., "The Risks of Solar Geoengineering Research," Science 372, no. 6547 (June 11, 2021): 1161, https://www.science.org/doi/10.1126/science.abj3679.

14. Edward A. Parson, "Geoengineering: Symmetric Precaution," Science 374, no. 6569 (November 11, 2021): 795, https://www.science.org/doi/10.1126/science.abm8462.

15. Holly Jean Buck, "Environmental Peacebuilding and Solar Geoengineering," Frontiers in Climate 4 (April 26, 2022), https://www.frontiersin.org/articles/10.3389/fclim.2022.869774/full.

16. David Keith, "What's the Least Bad Way to Cool the Planet?" New York Times, October 1, 2021, https://www.nytimes.com/2021/10/01/opinion/climate-change-geoengineering.html.

17. James Temple, "A Startup Says It's Begun Releasing Particles into the Atmosphere, in an Effort to Tweak the Climate," MIT Technology Review, December 24, 2022, https://www.technologyreview.com/2022

/12/24/1066041/a-startup-says-its-begun-releasing-particles-into-the
-atmosphere-in-an-effort-to-tweak-the-climate/.

Chapter 12: Meet the Neighbors

1. David Crookes, "Roswell UFO Crash: What Is the Truth Behind the 'Flying Saucer' Incident?" *All About Space*, May 6, 2021, https://www.livescience.com/roswell-ufo-crash-what-really-happened.html.
2. Mike Wall, "UFO Sightings Remain Mysterious, US Government Report Says," Space.com, June 26, 2021, https://www.space.com/us-government-ufo-report-released.
3. To read the US government's Office of the Director of National Intelligence report, "Preliminary Assessment: Unidentified Aerial Phenomena," June 25, 2021, visit https://www.dni.gov/files/ODNI/documents/assessments/Prelimary-Assessment-UAP-20210625.pdf.
4. If you want to see some pretty cool debunking of these videos (and others), check out Mick West's website, https://www.metabunk.org/home/.
5. "The Drake Equation Revisited: Part 1," *Astrobiology Magazine*, September 29, 2003, https://web.archive.org/web/20210225062139/http://www.astrobio.net/alien-life/the-drake-equation-revisited-part-i/.
6. Eric Betz, "The Wow! Signal: An Alien Missed Connection?" *Astronomy*, September 30, 2020, https://astronomy.com/news/2020/09/the-wow-signal-an-alien-missed-connection.
7. Ben Turner, "China Says It May Have Received Signals from Aliens," *Live Science*, June 15, 2022, https://www.livescience.com/china-says-it-may-have-received-signals-from-aliens.
8. Adolf Schaller, "Creating Life on a Gas Giant," Planetary Society, November 2, 2013, https://www.planetary.org/articles/20131023-on-hunters-floaters-and-sinkers-from-cosmos.
9. Dan Falk, "Why Extraterrestrial Life May Not Seem Entirely Alien," *Quanta Magazine*, March 18, 2021, https://www.quantamagazine.org/arik-kershenbaum-on-why-alien-life-may-be-like-life-on-earth-20210318/.
10. Andrew Lockley and Daniele Visioni, "Detection of Pre-industrial Societies on Exoplanets," *International Journal of Astrobiology* 20, no. 1 (February 2021): 73–80, https://www.cambridge.org/core/journals/international-journal-of-astrobiology/article/detection-of-preindustrial-societies-on-exoplanets/2F1C14870F756707F4808D2045AAA80C.

11. For more on KIC 8462852 or "Tabby's Star," visit https://en.wikipedia.org /wiki/Tabby%27s_Star.

12. Freeman J. Dyson, "Search for Artificial Stellar Sources of Infrared Radiation," *Science* 131, no. 3414 (June 3, 1960): 1667–68, https://www .science.org/doi/10.1126/science.131.3414.1667.

Part IV: Changed by Computing

Chapter 13: Robots

1. For more information on Nabis's early robot/torture device, visit https:// en.wikipedia.org/wiki/Apega_of_Nabis.

2. Tatiana Bur, "Mechanical Miracles: Automata in Ancient Greek Religion" (master's thesis, University of Sydney, 2016), 72, https://ses.library.usyd .edu.au/bitstream/2123/15398/1/bur_tcd_thesis.pdf.

3. For more information on Karel Čapek's 1920 play *R.U.R.*, visit https://en .wikipedia.org/wiki/R.U.R.

4. For more information on Isaac Asimov's story "Liar!" visit https://en .wikipedia.org/wiki/Liar!_(short_story).

5. Mark Robert Anderson, "After 75 Years, Isaac Asimov's Three Laws of Robotics Need Updating," *The Conversation*, March 17, 2017, https: //theconversation.com/after-75-years-isaac-asimovs-three-laws-of -robotics-need-updating-74501.

6. Christoph Salge and Daniel Polani, "Empowerment as Replacement for the Three Laws of Robotics," *Frontiers in Robotics and AI* 4 (June 29, 2017), https://www.frontiersin.org/articles/10.3389/frobt .2017.00025/full.

7. Nick Bostrom, *Superintelligence: Paths, Dangers, Strategies* (London: Oxford University Press, 2014), Chapter 9: "The Control Problem."

8. Margaret Boden et al., "Principles of Robotics: Regulating Robots in the Real World," *Connection Science* 29, no. 2 (2017): 124–29.

9. The videos of Boston Dynamics' humanoid robots called "Atlas" working as an assistant or running an indoor parkour layout make the case better than I could. Visit https://www.youtube.com/watch?v=tF4DML7FIWk and https://www.youtube.com/watch?v=-e1_QhJ1EhQ&t=6s.

10. Elizabeth Pennisi, "This Drone Has Legs: Watch a Flying Robot Perch on Branches, Catch a Tennis Ball in Midair," *Science*, December 1, 2021,

https://www.science.org/content/article/drone-has-legs-watch-flying -robot-perch-branches-catch-tennis-ball-midair.

11. Cecilia Laschi, "Bioinspired Soft Robotics: Lessons from Marine Species for Robot Applications," Autonomous Robotics Research Centre, March 24, 2021, https://www.youtube.com/watch?v=4u3OsSxCyWs.

12. John Timmer, "Researchers Build a Swimming Robot That Works in the Mariana Trench," Ars Technica, March 3, 2021, https://arstechnica.com/ science/2021/03/researchers-build-a-swimming-robot-that-works-in-the -mariana-trench/.

13. Jari Kätsyri, Meeri Mäkäräinen, and Tapio Takala, "Testing the 'Uncanny Valley' Hypothesis in Semirealistic Computer-Animated Film Characters: An Empirical Evaluation of Natural Film Stimuli," International Journal of Human-Computer Studies 97 (January 2017): 149–61, https://www .sciencedirect.com/science/article/pii/S1071581916301227.

14. Karl F. MacDorman et al., "Too Real for Comfort? Uncanny Responses to Computer Generated Faces," Computers in Human Behavior 25, no. 3 (May 2009): 695–710, https://www.sciencedirect.com/science/article/abs /pii/S0747563208002379.

15. For more information on Hanson Robotics' humanlike robot Sophia, visit https://www.hansonrobotics.com/sophia/.

16. Robert Sparrow, "The March of the Robot Dogs," Ethics and Information Technology 4 (December 2002): 305–18, https://link.springer.com/article /10.1023/A:1021386708994.

17. Todd Leopold, "HitchBOT, the Hitchhiking Robot, Gets Beheaded in Philadelphia," CNN, August 4, 2015, https://www.cnn.com/2015/08/03/ us/hitchbot-robot-beheaded-philadelphia-feat/index.html.

18. Christoph Bartneck and Merel Keijsers, "The Morality of Abusing a Robot," Paladyn: Journal of Behavioral Robotics, June 17, 2020, https: //www.degruyter.com/document/doi/10.1515/pjbr-2020-0017/ html?lang=en.

Chapter 14: AI Today

1. Corrina Schlombs, "How Ada Lovelace Used Her Knowledge of Music and Embroidery to Become a Computing Legend," Fast Company, December 13, 2022, https://www.fastcompany.com/90823737/how-ada -lovelace-used-her-knowledge-of-music-and-embroidery-to-become-a -computing-legend.

2. Geoffrey Jefferson, "The Mind of Mechanical Man," *British Medical Journal* 1, no. 4616 (January 25, 1949): 1105–10, https://www.bmj.com/content/1/4616/1105.

3. A. M. Turing, "Intelligent Machinery, A Heretical Theory," *Philosophia Mathematica* 4, no. 3 (1996): 256–60, https://rauterberg.employee.id.tue.nl/lecturenotes/DDM110%20CAS/Turing/Turing-1951%20Intelligent%20Machinery-a%20Heretical%20Theory.pdf.

4. "Unleashing the Promise of Artificial Intelligence in Radiology," GE Heathcare, September 2, 2021, https://www.gehealthcare.com/insights/article/unleashing-the-promise-of-artificial-intelligence-in-radiology.

5. For more information on CardiAI, visit https://cardiai.com/.

6. Ann Kelleher, "Moore's Law—Now and in the Future," Intel Newsroom, February 16, 2022, https://www.intel.com/content/www/us/en/newsroom/opinion/moore-law-now-and-in-the-future.html?wapkw=moore%27s%20law#gs.srry5r.

7. Olga Akselrod, "How Artificial Intelligence Can Deepen Racial and Economic Inequities," ACLU Newsroom, July 13, 2021, https://www.aclu.org/news/privacy-technology/how-artificial-intelligence-can-deepen-racial-and-economic-inequities.

8. Deborah Yao, "25 Years Ago Today: How Deep Blue vs. Kasparov Changed AI Forever," *AI Business*, May 11, 2022, https://aibusiness.com/ml/25-years-ago-today-how-deep-blue-vs-kasparov-changed-ai-forever.

9. Garry Kasparov, "Chess, a *Drosphilia* of Reasoning," *Science*, December 7, 2018, https://www.sciencemagazinedigital.org/sciencemagazine/07_december_2018/MobilePagedArticle.action?articleId=1447368.

10. Liz O'Sullivan and John Dickerson, "Here Are a Few Ways GPT-3 Can Go Wrong," *Tech Crunch*, August 7, 2020, https://techcrunch.com/2020/08/07/here-are-a-few-ways-gpt-3-can-go-wrong/.

11. Curtis Honeycutt, "Grammar Guy: Newspaper Headlines Obscure Writers' Views," *Savannah Now*, September 2, 2021, https://www.savannahnow.com/story/opinion/2021/09/02/newspaper-headlines-often-misrepresent-story-they-meant-help-tell/5647709001/.

12. Jefferson, "The Mind of Mechanical Man."

13. Quote from Jeremie Harris from a personal communication by phone January 26, 2023.

14. "A Skeptical Take on the AI Revolution," *The Ezra Klein Show*

podcast, January 6, 2023, https://www.nytimes.com/2023/01/06/podcasts /transcript-ezra-klein-interviews-gary-marcus.html.

15. James Vincent, "Nick Cave Says Imitation ChatGPT Song Is 'a Grotesque Mockery of What It Is to Be Human,'" *Verge*, January 17, 2023, https:// www.theverge.com/2023/1/17/23558572/nick-cave-chatgpt-imitation -lyrics-grotesque-mockery.

16. Annie Lowry, "How ChatGPT Will Destabilize White-Collar Work," *Atlantic*, January 20, 2023, https://www.theatlantic.com/ideas/archive /2023/01/chatgpt-ai-economy-automation-jobs/672767/.

Chapter 15: AI Tomorrow

1. Irving John Good, "Speculations Concerning the First Ultraintelligent Machine," *Advances in Computers* 6 (1966): 31–88, https://www.science direct.com/science/article/pii/S0065245808604180.

2. Nick Bostrom, *Superintelligence: Paths, Dangers, Strategies* (London: Oxford University Press, 2014), Chapter 8: "Is the Default Outcome Doom?"

3. "Why We Need to Rethink the Purpose of AI: A Conversation with Stuart Russell," *McKinsey on AI* podcast, May 12, 2020, https://www.mckinsey .com/capabilities/quantumblack/our-insights/why-we-need-to-rethink -the-purpose-of-ai-a-conversation-with-stuart-russell.

4. "Summary of Hans Moravec's *Robot: Mere Machine to Transcendent Mind*," *Reason and Meaning*, February 5, 2016, https://reasonandmeaning .com/2016/02/05/summary-of-hans-moravecs-robot-mere-machine-to -transcendent-mind/.

5. Charles Platt, "Superhumanism," *Wired*, October 1, 1995, https://www .wired.com/1995/10/moravec/.

6. Kelly Servick, "In a 'Tour de Force,' Researchers Image an Entire Fly Brain in Minute Detail," *Science*, July 19, 2018, https://www.science.org/content /article/tour-de-force-researchers-image-entire-fly-brain-minute-detail.

7. A. M. Turing, "Computing Machinery and Intelligence," *Mind* 59, no. 236 (October 1950): 433–60, https://academic.oup.com/mind/article/LIX /236/433/986238.

8. "Artificial Intelligence Alarmists Win ITIF's Annual Luddite Award," Information Technology & Information Foundation, January 19, 2016, https://itif.org/publications/2016/01/19/artificial-intelligence-alarmists -win-itif%E2%80%99s-annual-luddite-award/.

9. Kevin Kelly, "The Myth of a Superhuman AI," *Wired*, April 25, 2017, https://www.wired.com/2017/04/the-myth-of-a-superhuman-ai/.

10. Max Tegmark, "Friendly AI: Aligning Goals," Future of Life Institute, August 29, 2017, https://futureoflife.org/recent-news/friendly-ai-aligning-goals/.

11. Sean Carroll, "Melanie Mitchell on Artificial Intelligence and the Challenge of Common Sense," *Mindscape* podcast, Episode 68, October 14, 2019, https://www.preposterousuniverse.com/podcast/2019/10/14/68-melanie-mitchell-on-artificial-intelligence-and-the-challenge-of-common-sense/.

12. Stuart Russell, *Human Compatible: AI and the Problem of Control* (New York: Penguin Books, 2019), Chapter 7: "AI: A Different Approach."

Chapter 16: AI Gone Wild

1. Anil K. Seth and Tim Bayne, "Theories of Consciousness," *Nature Reviews Neuroscience* 23 (2022): 439–52.

2. Nitasha Kiku, "The Google Engineer Who Thinks the Company's AI Has Come to Life," *Washington Post*, June 11, 2022, https://www.washingtonpost.com/technology/2022/06/11/google-ai-lamda-blake-lemoine/.

3. Eric Schwitzgebel and Mara Garza, "A Defense of the Rights of Artificial Intelligences," University of California, Riverside, September 15, 2015, http://www.faculty.ucr.edu/~eschwitz/SchwitzPapers/AIRights-150915.pdf.

4. Julian Huxley, "Transhumanism," *Journal of Humanistic Psychology* 8, no. 1 (January 1968): 73–76, https://www.researchgate.net/publication/247718617_Transhumanism.

5. For more on Alcor, visit https://www.alcor.org/.

6. Melanie Mitchell, "Why AI Is Harder than We Think," Gecco '21: Proceedings of the Genetic and Evolutionary Computation Conference, June 2021, https://www.arxiv-vanity.com/papers/2104.12871/.

7. Nick Bostrom, "Are We Living in a Computer Simulation?" *Philosophical Quarterly* 53, no. 211 (2003): 243–55, https://philpapers.org/rec/BOSAWL.

8. Jason Kehe, "Of Course We're Living in a Simulation," *Wired*, March 9, 2022, https://www.wired.com/story/living-in-a-simulation/.

9. "Dartmouth Physicist Slams *Matrix* Idea That Life Is an Aliens' Sim," *Mind Matters*, July 10, 2022, https://mindmatters.ai/2022/07/dartmouth-physicist-slams-matrix-idea-that-life-is-an-aliens-sim/.

10. John D. Barrow, "Living in a Simulated Universe," from *Universe or Multiverse?*, edited by Bernard Carr (Cambridge: Cambridge University Press, 2007), 481–86, https://www.simulation-argument.com/barrowsim.pdf.

11. Fouad Khan, "Confirmed! We Live in a Simulation," *Scientific American*, April 1, 2021, https://www.scientificamerican.com/article/confirmed-we-live-in-a-simulation/.

12. "Life's a Sim and Then You're Deleted," *New Scientist*, July 27, 2002, https://www.newscientist.com/article/mg17523535-900-lifes-a-sim-and-then-youre-deleted/.

13. Jess Keiser, "Our World Might Be a Simulation. Would That Be So Bad?" *Washington Post*, February 11, 2022, https://www.washingtonpost.com/outlook/2022/02/11/chalmers-reality-review-simulation-hypothesis-philosophy/.

Index

National Space Society, 150
natural killer cells, 136
natural selection, 14, 19
"Natural versus Urban Scenes: Some
	Psychophysiological Effects" (Ulrich),
	133–34
Nature (journal), 46
nature, human need for, 133–36
Nature Communications, 17
Nature Energy, 242
Neanderthals, 10, 20, 144n, 185
negative-emission technologies, 160
nervous system, 37, 192
Netflix, 202n
Netherlands, 73, 190
net-zero emissions, 154
neural dust, 41–43
Neuralink, 42–43
neural lace, 42
neural networks, 50
Neuron (journal), 2
neurons, 2–3, 221
newborns, 50
New Mexico, 159
Newsweek, 21, 24
New York City, 77
New York Times, 166
New York University, 211, 244
New Zealand, 196
Nietzsche, Friedrich, 234
Nippon Medical School, 136
nitrogen, 80, 90, 235
nitrogen oxide, 179
Nobel Peace Prize, 72
noise pollution, 111
nongovernmental organizations (NGOs),
	155
Northeastern University, 164, 165
Northwest Territories, 142
Norway, 103
Novak, Ben, 144–45
Novameat, 84
Novoselov, Valery, 57
nuclear fusion, 1
nuclear holocaust, 148, 153

nuclear war, 71
nutrition, 13
Nysa, 187

Oak Ridge National Laboratory, 130
obsessive-compulsive disorder, 43
Occidental Petroleum, 159
octopi, 192, 223
Oculus helmet, 146
offices, commuting to, 109
Ohio River, 144
Ohio State University, 175
Okin, Gregory, 86
Olympic Games, 31
O'Neill, Gerard, 149–50
O'Neill colonies, 149, 150n
1.5-degrees Celsius barrier, 154, 162
1PointFive (company), 159
OpenAI, 207, 210
opioid crisis, 57
orbital flights, 114–15
Orca, 158, 159, 166
organ donors, 22
organ transplants, 21–25, 62
Outer Limits, The (TV show), 9–10, 16–17
Overture, 111
Oxford University, 53, 239

Pager (monkey), 43
pain, chronic, 43
Panama disease Race 1 (fungus), 95, 96
pangenome, 47
parabiosis, 63, 64n
parakeets, 142
Paralympic Games, 31
paralysis, 40–41
Parkinson's disease, 43
parrots, 228
Parson, Edward, 165
passenger pigeon, 101, 142, 144, 145
passive internal airflow, 132
pasteurization, 56
PayPal, 64
Pearce, Mick, 132
Pedestrian Observations (blog), 117

About the Author

Jay Ingram has hosted two national science programs in Canada, *Quirks & Quarks* on CBC Radio and *Daily Planet* on Discovery Channel Canada. He is the author of twenty books, which have been translated into fifteen languages, including the bestselling five-volume *The Science of Why* series. In 2015 he won the Walter C. Alvarez Award from the American Medical Writers Association for excellence in communicating health care developments and concepts to the public, and from 2005 to 2015 he chaired the Science Communications Program at the Banff Centre. Jay has seven honorary degrees, was awarded the Queen Elizabeth II Diamond Jubilee Medal, and is a Member of the Order of Canada. He is cofounder of the arts and engineering street festival called Beakerhead, in Calgary, Alberta. He lives in Calgary. Connect with him at JayIngram.ca or on X @JayIngram.